ANNOUNCING...

Math Contests
for
High School
Volume 8

School Years: 2016-2017 through 2020-2021

Written by

Steven R. Conrad • Daniel Flegler • Adam Raichel

Published by MATH LEAGUE PRESS
Printed in the United States of America

Cover art by Bob DeRosa

First Printing, 2021
Copyright © 2021
by Mathematics Leagues Inc.
All Rights Reserved

Math League Press
P.O. Box 17
Tenafly, NJ 07670-0017

ISBN 978-0-940805-26-2

Preface

Math Contests—High School, Volume 8 is the eighth volume in our series of problem books for high school students. The first seven volumes contain contests given in the school years 1977-1978 through 2015-2016. Volume 8 contains contests given from 2016-2017 through 2020-2021. (Use the order form on page 70 to order any of our 24 books.)

These books give classes, clubs, teams, and individuals diversified collections of high school math problems. All of these contests were used in regional interscholastic competition throughout the United States and Canada. Each contest was taken by about 80 000 students. In the contest section, each page contains a complete contest that can be worked during a 30-minute period. The convenient format makes this book easy to use in a class, a math club, or for just plain fun. In addition, detailed solutions for each contest also appear on a single page.

Every contest has questions from different areas of mathematics. The goal is to encourage interest in mathematics through solving *worthwhile* problems. Many students first develop an interest in mathematics through problem-solving activities such as these contests. On each contest, the last two questions are generally more difficult than the first four. The final question on each contest is intended to challenge the very best mathematics students. The problems require no knowledge beyond secondary school mathematics. No knowledge of calculus is required to solve any of these problems. From two to four questions on each contest are accessible to students with only a knowledge of elementary algebra. Starting with the 1992-93 school year, students have been permitted to use any calculator without a QWERTY keyboard on any of our contests.

This book is divided into four sections for ease of use by both students and teachers. The first section of the book contains the contests. Each contest contains six questions that can be worked in a 30-minute period. The second section of the book contains detailed solutions to all the contests. Often, several solutions are given for a problem. Where appropriate, notes about interesting aspects of a problem are mentioned on the solutions page. The third section of the book consists of a listing of the answers to each contest question. The last section of the book contains the difficulty rating percentages for each question. These percentages (based on actual student performance on these contests) determine the relative difficulty of each question.

You may prefer to consult the answer section rather than the solution section when first reviewing a contest. The authors believe that reworking a problem, knowing the answer (but *not* the solution), often helps to better understand problem-solving techniques.

Revisions have been made to the wording of some problems for the sake of clarity and correctness. The authors welcome comments you may have about either the questions or the solutions. Though we believe there are no errors in this book, each of us agrees to blame the others should any errors be found!

Steven R. Conrad, Daniel Flegler, & Adam Raichel, contest authors

Acknowledgments

For the beauty, cleverness, and breadth of his numerous mathematical contributions for the past 40 years, we are indebted to Michael Selby.

For her continued patience and understanding, special thanks to Marina Conrad, whose only mathematical skill, an important one, is the ability to count the ways.

For demonstrating the meaning of selflessness on a daily basis, special thanks to Grace Flegler.

To Daniel Will-Harris, whose skill in graphic design is exceeded only by his skill in writing *really* funny computer books, thanks for help when we needed it most: the year we first began to typeset these contests on a computer.

Table Of Contents

School Year	Contest #	Page # for Problems	Page # for Solutions	Page # for Answers	Page # for Difficulty Ratings
2016-2017	1	2	34	66	68
2016-2017	2	3	35	66	68
2016-2017	3	4	36	66	68
2016-2017	4	5	37	66	68
2016-2017	5	6	38	66	68
2016-2017	6	7	39	66	68
2017-2018	1	8	40	66	68
2017-2018	2	9	41	66	68
2017-2018	3	10	42	66	68
2017-2018	4	11	43	66	68
2017-2018	5	12	44	66	68
2017-2018	6	13	45	66	68
2018-2019	1	14	46	66	68
2018-2019	2	15	47	66	68
2018-2019	3	16	48	66	68
2018-2019	4	17	49	66	68
2018-2019	5	18	50	66	68
2018-2019	6	19	51	66	68
2019-2020	1	20	52	67	68
2019-2020	2	21	53	67	68
2019-2020	3	22	54	67	68
2019-2020	4	23	55	67	68
2019-2020	5	24	56	67	68
2019-2020	6	25	57	67	68
2020-2021	1	26	58	67	68
2020-2021	2	27	59	67	68
2020-2021	3	28	60	67	68
2020-2021	4	29	61	67	68
2020-2021	5	30	62	67	68
2020-2021	6	31	63	67	68

The Contests

October, 2016 – March, 2021

HIGH SCHOOL MATHEMATICS CONTESTS

Math League Press, P.O. Box 17, Tenafly, New Jersey 07670-0017

Contest Number 1 *Any calculator without a QWERTY keyboard is allowed.* Answers must be exact *or* have 4 (or more) significant digits, correctly rounded. **October 18, 2016**

Name _____ Teacher _____ Grade Level _____ Score _____

Time Limit: 30 minutes | *Answer Column*

1-1. What is the greatest common divisor of 2016 and 2017? | 1-1.

1-2. If the 3-digit number 1A7 (where A is the tens digit) is divisible by 3, but not by 9, what are all possible values of A? | 1-2.

1-3. When a small square is surrounded by 4 congruent rectangles in the manner shown, a larger square is formed. If the perimeter of each rectangle is 18, what is the area of the larger square? | 1-3.

1-4. The length of one leg of right $\triangle T$ is the average of the lengths of the other two sides. If the perimeter of T is 1, what is T's area? | 1-4.

1-5. At my grandparents' dance party, each guest ate at least 1 almond. If each guest ate a whole number of almonds, no three guests ate the same number of almonds, and the guests ate at most 2600 almonds all together, then at most how many guests were at the dance party? | 1-5.

1-6. In the five-term sequence 60, 20, 30, 40, 50 the least term is 20, the greatest term is 60, and the nth term is divisible by n. What is the least possible sum of all the terms of an eleven-term sequence of unequal positive integers whose least term is 12 and whose nth term is divisible by n? | 1-6.

Contest Number 2 *Any calculator without a QWERTY keyboard is allowed.* Answers must be exact *or* have 4 (or more) significant digits, correctly rounded. **November 15, 2016**

Name _____ Teacher _____ Grade Level _____ Score _____

Time Limit: 30 minutes   *Answer Column*

2-1. My dog Fifi is more than 1 year old. If she were 3 years older, her age would be the square of a certain integer. If she were 3 years younger, then her age would be that certain integer. How old is my dog?

2-1.

2-2. A non-square rhombus of side-length 5 is inscribed in a rectangle as shown (but the diagram is NOT drawn to scale). If every side of the rectangle has an integral length, what is the least possible area of the rectangle?

2-2.

2-3 What is the sum of the real numbers x and y which satisfy both

$$1159x + 857y = 2798$$
$$\text{and } 857x + 1159y = 1234?$$

2-3

2-4. If $k = 534^{620}$, what is the units digit of 3^k?

2-4.

2-5 If $f(x) = 3f(1-x) + 1$ for all x, what is the value of $f(2016)$?

2-5

2-6. What is the area of a circle in which the lengths of two adjacent sides of an inscribed hexagon are each 2, and the lengths of the remaining four sides of the hexagon are each 12?

2-6.

HIGH SCHOOL MATHEMATICS CONTESTS

Math League Press, P.O. Box 17, Tenafly, New Jersey 07670-0017

Contest Number 3 *Any calculator without a QWERTY keyboard is allowed.* Answers must be exact *or* have 4 (or more) significant digits, correctly rounded. **December 13, 2016**

Name _____ Teacher _____ Grade Level _____ Score _____

Time Limit: 30 minutes

Answer Column

3-1.   For what digit A will the 6-digit number $72871A$ be divisible by 7?

3-1. _____

3-2.   One year, the month of January had exactly 4 Mondays and exactly 4 Fridays. On what day of the week did January 1 fall that year?

3-2. _____

3-3.   The first 1000 positive integers are written consecutively. Starting at 15, every multiple of 15 is circled, Starting at 21, every multiple of 21 is circled. What is the least *positive* difference between two unequal circled numbers?

3-3. _____

3-4.   When n is divided by 2016, the remainder is 1008. When $2n$ is divided by 2016, the remainder is k. For what integer k is this true?

3-4. _____

3-5.   As seen at the right, an 8×20 rectangle drawn horizontally intersects a 6×28 rectangle at an acute angle of 45°. What is the area of the shaded region that is common to the interiors of these two rectangles?

3-5. _____

3-6.   What is the largest integer n for which a circle with center $(\sqrt{2}, \sqrt{2})$ can pass through exactly n points with integer coordinates?

3-6. _____

Solutions on Page 36 • Answers on Page 66

HIGH SCHOOL MATHEMATICS CONTESTS

Math League Press, P.O. Box 17, Tenafly, New Jersey 07670-0017

Contest Number 4 *Any calculator without a QWERTY keyboard is allowed.* Answers must be exact *or* have 4 (or more) significant digits, correctly rounded. **·January 10, 2017**

Name _____ Teacher _____ Grade Level _____ Score _____

Time Limit: 30 minutes | *Answer Column*

4-1. How much greater is the sum of the first 2017 positive even integers than the sum of the first 2017 positive odd integers? | 4-1.

4-2. Define the *divisor-sum* of a positive integer n as the sum of n's positive integer divisors. For example, the divisor-sum of 2 is a prime because $1 + 2 = 3$, and 3 is a prime. What is the least positive integer greater than 2 whose divisor-sum is a prime? | 4-2.

4-3. If we get a bullseye half the time that we play darts, is it most likely that we **A)** miss a bullseye in any single try, **B)** get a bullseye at least 1 time in 2 tries, or **C)** get a bullseye 2 or 3 times in 3 tries? | 4-3.

4-4. Two externally tangent circles are both tangent to large circle C at opposite ends of a diameter of C. A chord of C, tangent to both small circles, has length 8. The diameter of the larger unshaded circle is 6. What is the area of the shaded part of C? | 4-4.

4-5. What is the least real number a for which there is at least one real solution x of the equation $(\log_{10}x)^2 + 6\log_{10}x = a$? | 4-5.

4-6. If $S = \dfrac{4^{\frac{1}{2000}}}{4^{\frac{1}{2000}}+2} + \dfrac{4^{\frac{2}{2000}}}{4^{\frac{2}{2000}}+2} + \dfrac{4^{\frac{3}{2000}}}{4^{\frac{3}{2000}}+2} + \ldots + \dfrac{4^{\frac{1998}{2000}}}{4^{\frac{1998}{2000}}+2} + \dfrac{4^{\frac{1999}{2000}}}{4^{\frac{1999}{2000}}+2}$, | 4-6.

write the value of S as a decimal in standard form.

© 2017 by Mathematics Leagues Inc.

HIGH SCHOOL MATHEMATICS CONTESTS

Math League Press, P.O. Box 17, Tenafly, New Jersey 07670-0017

Contest Number 5   *Any calculator without a QWERTY keyboard is allowed.* Answers must be exact *or* have 4 (or more) significant digits, correctly rounded.   **February 7, 2017**

Name _____ Teacher _____ Grade Level ____ Score ____

Time Limit: 30 minutes   *Answer Column*

5-1. If the squares of the lengths of two sides of a right triangle are 16 and 20, what are both possible lengths of the third side?

5-1.

5-2. What integer between 2 and 500 is the square of an integer but has none of the numbers 2, 3, 5, 7, 11, 13, or 17 as a factor?

5-2.

5-3. If a and b are nonzero real numbers that satisfy $\frac{b}{1+ab} = 2017$, what is the value of $a + \frac{1}{b}$?

5-3.

5-4. If $z^{12} = 1$ and $z^{20} = 1$, what are all possible values, real or imaginary, of the complex number z?

5-4.

5-5. Two large right triangles share a hypotenuse of length 17. The triangles cross where one of the legs is split into segments of length 6 and 9, as shown. What is the area of the shaded large right triangle?

5-5.

5-6. How many different ordered pairs of positive integers (m,n) satisfy
$$\frac{1}{m} + \frac{1}{n} + \frac{1}{mn} = \frac{1}{5}?$$

5-6.

Solutions on Page 38 • Answers on Page 66

HIGH SCHOOL MATHEMATICS CONTESTS

Math League Press, P.O. Box 17, Tenafly, New Jersey 07670-0017

Contest Number 6   *Any calculator without a QWERTY keyboard is allowed.* Answers must be exact *or* have 4 (or more) significant digits, correctly rounded.   **March 14, 2017**

Name _____ Teacher _____ Grade Level _____ Score ____

Time Limit: 30 minutes   *Answer Column*

6-1.   If $(x-2016)(2018-x) = 1$, what is the value of $(x-2018)(2016-x)$? | 6-1.

6-2.   On some Fridays, twins You and Me cry together, and on some Fridays they don't. Half the time, at least one of them cries on a Friday. One-eighth of the time, Me cries on a Friday and You doesn't. What is the probability that You cries on a Friday? | 6-2.

6-3.   How many different ordered pairs of real numbers (a,b) satisfy
$$a^2 + b^2 = ab?$$ | 6-3.

6-4.   Two parallel diagonals of a regular hexagon split the hexagon into three regions, as shown. What is the ratio of the sum of the areas of the two shaded regions to the area of the regular hexagon? | 6-4.

6-5.   The length of the shorter leg of right triangle T is 2. If the bisector of T's larger acute angle intersects T's longer leg 1 unit from the vertex of T's right angle, how long is T's longer leg? | 6-5.

6-6.   In an increasing sequence of 50 different integers, each greater than 1, the difference between consecutive terms is constant. If the first term has no factor greater than 1 in common with any other term, what is the least possible value of the sequence's last term? | 6-6.

HIGH SCHOOL MATHEMATICS CONTESTS

Math League Press, P.O. Box 17, Tenafly, New Jersey 07670-0017

Contest Number 1 *Any calculator without a QWERTY keyboard is allowed.* Answers must be exact *or* have 4 (or more) significant digits, correctly rounded. **October 17, 2017**

Name _____ Teacher _____ Grade Level _____ Score ____

Time Limit: 30 minutes | *Answer Column*

1-1. If $\dfrac{1}{x + 2017} = 1$, what is the value of $\dfrac{1}{x + 2018}$? | 1-1.

1-2. Each pirate wants his own treasure chest, but there is 1 more pirate than there are treasure chests. If the pirates would agree to pair up so each pirate shares a treasure chest with another pirate, then 1 treasure chest would not be assigned to any pirate. How many treasure chests are there? | 1-2.

1-3. What is the greatest possible perimeter of a rectangle whose length and width are different prime numbers, each less than 120? | 1-3.

1-4. A rectangle is partitioned into 9 different squares, as shown at the right. The area of the smallest square, shown fully darkened, is 1. Two other squares have areas of 196 and 324, as shown. What is the area of the shaded square? | 1-4.

| 196 | 324 |

1-5. In a regular 10-sided polygon, two pairs of different vertices (four different vertices altogether) are chosen at random, so that all points chosen are distinct from each other. What is the probability that the line segments determined by each pair of points do *not* intersect? | 1-5.

1-6. What is the largest integer N for which $7x + 11y = N$ has no solution in non-negative integers (x,y)? | 1-6.

HIGH SCHOOL MATHEMATICS CONTESTS

Math League Press, P.O. Box 17, Tenafly, New Jersey 07670-0017

Contest Number 2 *Any calculator without a QWERTY keyboard is allowed.* Answers must be exact *or* have 4 (or more) significant digits, correctly rounded. **November 14, 2017**

Name _____ Teacher _____ Grade Level _____ Score _____

Time Limit: 30 minutes   *Answer Column*

2-1. What is the greatest value of x that satisfies $$(x+2017)(x+2018)(x+2019)(x+2020) = 0?$$	2-1.
2-2. If m and n are positive integers that satisfy $\sqrt{m} + \sqrt{n} = 10$, what is the greatest possible value of $m+n$?	2-2.
2-3. Mom, Dad, and I each write a positive integer. My number is least and Dad's is greatest. The average of all 3 numbers is 20. The average of the 2 smallest numbers is 8. If Dad's number is d and if my number is m, what is the greatest possible value of $d-m$?	2-3.
2-4. When the square of an eight-digit integer is subtracted from the square of a different eight-digit integer, sometimes the difference will be an eight-digit number with eight identical even digits. What are both possible values of the repeated digit in such a situation?	2-4.
2-5. A line segment is drawn from the upper right vertex of a parallelogram, as shown, dividing the opposite side into segments with lengths in a 2:1 ratio. If the area of the parallelogram is 90, what is the area of the shaded region?	2-5.
2-6. There are only two six-digit integers n greater than $100\,000$ for which n^2 has n as its final six digits (or, equivalently, for which n^2-n is divisible by 10^6). One of the integers is $890\,625$. What is the other?	2-6.

Contest Number 3 *Any calculator without a QWERTY keyboard is allowed.* Answers must be exact *or* have 4 (or more) significant digits, correctly rounded. **December 12, 2017**

Name _____ Teacher _____ Grade Level _____ Score _____

Time Limit: 30 minutes | *Answer Column*

3-1. What is the least possible value of a for which $a^2 + 40^2 = 50^2$? | 3-1.

3-2. There are an infinite number of points with positive coordinates (x,y) the sum of whose coordinates is the square of an integer. Among all such points (x,y), which one satisfies $y = 2x$ and has x as small as possible? | 3-2.

3-3. If 8 different integers are chosen at random from the first 15 positive integers, what is the probability that an additional number chosen at random from the remaining 7 positive integers is smaller than every one of the 8 originally chosen positive integers? | 3-3.

3-4. If the perimeter of an isosceles triangle with integral sides is 2017, how many different lengths are possible for the legs? | 3-4.

3-5. If $0 < a \le b \le 1$, what is the maximum value of $ab^2 - a^2b$? | 3-5.

3-6. A hexagon is inscribed in a circle as shown. If lengths of three sides of the hexagon are each 1 and the lengths of the other three sides are each 2, what is the area of this hexagon? | 3-6.

Contest Number 4 *Any calculator without a QWERTY keyboard is allowed. Answers must be exact or have 4 (or more) significant digits, correctly rounded.* **January 9, 2018**

Name _____ Teacher _____ Grade Level _____ Score _____

Time Limit: 30 minutes | *Answer Column*

4-1. What is the length of a side of the square whose area is numerically 2018 times its perimeter? | 4-1.

4-2. As shown, a small square is inscribed in one of the triangles formed when both diagonals of a larger square are drawn. If the area of the larger square is 144, what is the area of the smaller square? | 4-2.

4-3. What sequence of 5 positive integers has these three properties:
1) All but one of the numbers is a multiple of 5.
2) Every number after the first is 1 more than the sum of all the preceding numbers.
3) The first number is as small as possible. | 4-3.

4-4. What are all ordered triples of positive primes (p, q, r) which satisfy
$$p^q + 1 = r?$$ | 4-4.

4-5. What are all ordered pairs of integers (x, y) that satisfy
$$5x^3 + 2xy - 23 = 0?$$ | 4-5.

4-6. If x is a number chosen uniformly at random between 0 and 1, what is the probability that the greatest integer $\leq \log_2\left(\frac{1}{x}\right)$ is odd? | 4-6.

© 2018 by Mathematics Leagues Inc.

HIGH SCHOOL MATHEMATICS CONTESTS

Math League Press, P.O. Box 17, Tenafly, New Jersey 07670-0017

Contest Number 5 *Any calculator without a QWERTY keyboard is allowed.* Answers must be exact *or* have 4 (or more) significant digits, correctly rounded. **February 13, 2018**

Name _____ Teacher _____ Grade Level _____ Score ____

Time Limit: 30 minutes   *Answer Column*

5-1. What is the greatest possible product of 2 primes whose sum is 30?	5-1.
5-2. Trisection points on opposite sides of a rectangle are joined, as shown. If the area of the shaded region is 2018, what is the area of the rectangle?	5-2.
5-3. Three beavers (one not shown) take turns biting a tree until it falls. The second beaver is twice as likely as the first to make the tree fall. The third is twice as likely as the second to make the tree fall. What is the probability that a bite taken by the third beaver causes the tree to fall?	5-3.
5-4. The reflection of (6,3) across the line $x = 4$ is (2,3). If $m \neq 4$, what is the reflection of (m,n) across the line $x = 4$?	5-4.
5-5. If two altitudes of a triangle have lengths 10 and 15, what is the smallest integer that could be the length of the third altitude?	5-5.
5-6. In the interval $-1 < x < 1$, $\sin\theta$ is one root of $x^4 - 4x^3 + 2x^2 - 4x + 1 = 0$. In that same interval, for what ordered pair of integers (a,b) is $\cos 2\theta$ one root of $x^2 + ax + b = 0$?	5-6.

Contest Number 6 *Any calculator without a QWERTY keyboard is allowed.* Answers must be exact *or* have 4 (or more) significant digits, correctly rounded. **March 20, 2018**

Name _____ Teacher _____ Grade Level _____ Score _____

Time Limit: 30 minutes　　　FINAL CONTEST OF THE YEAR　　　*Answer Column*

6-1. What is the product of all four positive integer divisors of 2018?

6-1.

6-2. A unit fraction is a fraction whose numerator is 1 and whose denominator is a positive integer. What is the largest rational number that can be written as the sum of 3 different unit fractions?

6-2.

6-3. What is the ratio, larger to smaller, of a rectangle's dimensions if half of the rectangle is similar to the original rectangle?

6-3.

6-4. The vertices of a triangle are (8,7), (0,1), and (8,1). What are the coordinates of all points inside this triangle that have integral coordinates and lie on the bisector of the smallest angle of the triangle?

6-4.

6-5. If h is the number of heads obtained when 4 fair coins are each tossed once, what is the expected (average) value of h^2?

6-5.

6-6. Let $P(x) = 2x^{10} + 3x^9 + 4x + 9$. If z is a non-real solution of $z^3 = 1$, what is the numerical value of $P\left(\frac{1}{z}\right) + P\left(\frac{1}{z^2}\right) + P\left(\frac{1}{z^3}\right)$?

6-6.

HIGH SCHOOL MATHEMATICS CONTESTS

Math League Press, P.O. Box 17, Tenafly, New Jersey 07670-0017

Contest Number 1 *Any calculator without a QWERTY keyboard is allowed.* Answers must be exact *or* have 4 (or more) significant digits, correctly rounded. **October 16, 2018**

Name _____ Teacher _____ Grade Level ____ Score ____

Time Limit: 30 minutes | *Answer Column*

1-1. If the sum of two consecutive even integers is 2018, what is their average? | 1-1.

1-2. What is the area of a rectangle whose shorter sides have length 3 if a line segment drawn from a vertex of the rectangle to the midpoint of the rectangle's longer side (as shown) has length 5? | 1-2.

1-3. At 12:30, what is the degree-measure of the smaller angle formed by the minute and hour hands of a circular 12-hour clock? | 1-3.

1-4. I have n red cards and n black cards. If $25 \le n \le 50$, for how many values of n can I split my $2n$ cards into n groups, each having exactly two cards, if exactly one of these n groups contains 1 red card and 1 black card? | 1-4.

1-5. How many different positive integers leave a remainder of 24 when divided into 9449? | 1-5.

1-6. If 5 boys and 2 girls are seated at random in a toboggan, what is the probability that 4 or more boys are seated in consecutive places? | 1-6.

Solutions on Page 46 • Answers on Page 66

HIGH SCHOOL MATHEMATICS CONTESTS

Math League Press, P.O. Box 17, Tenafly, New Jersey 07670-0017

Contest Number 2 *Any calculator without a QWERTY keyboard is allowed.* Answers must be exact *or* have 4 (or more) significant digits, correctly rounded. **November 13, 2018**

Name _____ Teacher _____ Grade Level _____ Score _____

Time Limit: 30 minutes

		Answer Column
2-1.	Numerically, what is the area of a square two of whose sides have lengths $x-1$ and $5-x$?	2-1.
2-2.	What is the sum of the two integers k for which $4x^2 + kxy + 4y^2$ is the square of a polynomial in x and y?	2-2.
2-3.	Two isosceles triangles with perimeters 18 and 28 share a common base and have their third vertices on the same side of that base, as shown. If the legs of the larger triangle are twice as long as those of the smaller, then how long is their common base?	2-3.
2-4.	What are all positive integers n for which the least common multiple of n and 1000 is 2000?	2-4.
2-5.	Jan replaced each of the 20 different numbers in sequence S with its rank (**smallest to largest**) within S: 1 for smallest, 2 for second smallest, etc. The sum of the ranks of Jan's first 5 numbers was 66. Ann replaced each number in S with its rank (**largest to smallest**) within S: 1 for largest, ..., 20 for smallest. What was the sum of the ranks of Ann's first 5 numbers?	2-5.
2-6.	What is the largest integer less than 2018 which cannot be written as the sum of two or more consecutive positive integers?	2-6.

Solutions on Page 47 • Answers on Page 66

15

HIGH SCHOOL MATHEMATICS CONTESTS

Math League Press, P.O. Box 17, Tenafly, New Jersey 07670-0017

Contest Number 3 *Any calculator without a QWERTY keyboard is allowed.* Answers must be exact *or* have 4 (or more) significant digits, correctly rounded. **December 11, 2018**

Name _____ Teacher _____ Grade Level ____ Score ____

Time Limit: 30 minutes | *Answer Column*

3-1. If $n = (1000001-1)(1000001+1)$, what is the smallest positive integer k for which $n + k$ is the square of an integer?

3-1.

3-2. If $x + y = \dfrac{1}{x} + \dfrac{1}{y}$ and $x + y \neq 0$, what is the value of xy?

3-2.

3-3. If 2018 kids each take 1 or 2 bites from the same hot dog, then at most how many kids can take more than the mean number of bites of all 2018 kids?

3-3.

3-4. A circle which is centered at $O(0,0)$ intersects the x-axis at points $P(1,0)$ and $Q(-1,0)$. Point A is on the circle in the first quadrant so that $m\angle AOP = 30°$. A line drawn through A and parallel to the x-axis intersects the circle at point B in the second quadrant. What is the degree-measure of $\angle BQO$?

3-4.

3-5. A sequence is *geometric* if the ratio of each term to its preceding term always has the same value. In a geometric sequence, if the sum of the first 3 terms is 3 and the sum of the first 6 terms is 4, what is the sum of the first 12 terms?

3-5.

3-6. As shown at the right, three line segments partition a regular hexagon into four regions. Each unshaded region is labeled with its area. What is the area of the shaded (fourth) region?

17 7 7

3-6.

Solutions on Page 48 • Answers on Page 66

HIGH SCHOOL MATHEMATICS CONTESTS

Math League Press, P.O. Box 17, Tenafly, New Jersey 07670-0017

Contest Number 4 *Any calculator without a QWERTY keyboard is allowed.* Answers must be exact *or* have 4 (or more) significant digits, correctly rounded. **January 8, 2019**

Name _____ Teacher _____ Grade Level _____ Score ____

Time Limit: 30 minutes | *Answer Column*

4-1. What are both values of x that satisfy $\sqrt{x^2} - 2019 = 0$? | 4-1.

4-2. How long is the hypotenuse of an isosceles right triangle whose area is 4? | 4-2.

4-3. A triangle is inscribed in a circle of area 100π. If the length of one side of this triangle is 10, and the center of the circle lies inside the triangle, what is the degree-measure of the angle of this triangle that is opposite this side? | 4-3.

4-4. If the average of the 25 terms of a sequence of consecutive integers equals the square of the first term, what is the least possible value of the last term? | 4-4.

4-5. If $3^a = 9^b = 27^c$, and if $abc \neq 0$, what is the value of $\frac{a}{b} + \frac{b}{c} + \frac{c}{a}$? | 4-5.

4-6. What is the least possible sum of 5 different positive integers whose reciprocals form an arithmetic sequence? [**NOTE:** If a, b, and c form an arithmetic sequence with $a < b < c$, then $c-b = b-a$.] | 4-6.

HIGH SCHOOL MATHEMATICS CONTESTS

Math League Press, P.O. Box 17, Tenafly, New Jersey 07670-0017

Contest Number 5 *Any calculator without a QWERTY keyboard is allowed.* Answers must be exact *or* have 4 (or more) significant digits, correctly rounded. **February 12, 2019**

Name _____ Teacher _____ Grade Level _____ Score ____

Time Limit: 30 minutes   *Answer Column*

5-1. If N is a 3-digit integer, and given that the result of reversing the digits of N is the number M, what is the maximum value of $N-M$? [Note: Integers written with 1 or more leading zeros can also be written with every leading zero removed.]

5-1.

5-2. If a and b are integers whose product is 5, what is the least possible value of a^b?

5-2.

5-3. Al's biplane can fly two different types of square advertising signs. Both signs have integer side-lengths. Their areas are $n+1$ and $2n+1$. What is the least positive integer n for which each sign's area is the square of an integer?

5-3.

5-4. In a group of 100, each a Borg or a Corg, every Borg has 2 Corg friends and no Corg has more than 1 Borg friend. If exactly 13 Corgs have no Borg friends, then how many Corgs are in the group?

5-4.

5-5. If 2019! is the product of the first 2019 positive integers, for which positive integer k is $x = 2019!$ the only value of $x > 0$ that satisfies

$$\frac{1}{\log_4 x} + \frac{1}{\log_9 x} + \frac{1}{\log_{16} x} + \ldots + \frac{1}{\log_{(2019^2)} x} = k?$$

5-5.

5-6. Midpoints of two sides of a square are vertices of the shaded triangle shown. Drawing the diagonal of the square pictured splits this triangle into two parts, one of which is a trapezoid. If the area of the square is 192, what is the area of the trapezoid?

5-6.

HIGH SCHOOL MATHEMATICS CONTESTS

Math League Press, P.O. Box 17, Tenafly, New Jersey 07670-0017

Contest Number 6 *Any calculator without a QWERTY keyboard is allowed.* Answers must be exact *or* have 4 (or more) significant digits, correctly rounded. **March 19, 2019**

Name _____ Teacher _____ Grade Level _____ Score ____

Time Limit: 30 minutes

Answer Column

6-1. A square is partitioned into three congruent rectangles, as shown. If the perimeter of one of these rectangles is 40, what is the perimeter of the square?

6-1.

6-2. What is the smallest integer greater than 237 for which each digit of the integer is a prime number?

6-2.

6-3. Each of the 16 integers 0, 1, 2, . . . , 13, 14, 15 is placed in 1 of 4 sets A, B, C, or D so that each set has 4 different numbers. The sum of the numbers in set B is 2 times the sum of the numbers in set A. The sum of the numbers in set C is 4 times the sum of the numbers in set A. The sum of the numbers in set D is 5 times the sum of the numbers in set A. What is the sum of the numbers in set A?

6-3.

6-4. A line segment has one endpoint at $(a,0)$ and its other endpoint on the line $y = x$. If the midpoint of this line segment is $(2019,2020)$ what is the value of a?

6-4.

6-5. Of the (radian) real number pairs (x,y) for which $\cos x \cos y = \frac{3}{4}$ and $\sin x \sin y = -\frac{1}{4}$, which two pairs minimize both $|x|$ and $|y|$?

6-5.

6-6. Choose 10 points on a circle so that no 3 chords with endpoints among these 10 points are ever concurrent interior to the circle. If chords join all pairs of these points, how many different triangles are formed that have their sides on these chords and their vertices interior to the circle at points of intersection of these chords?

6-6.

HIGH SCHOOL MATHEMATICS CONTESTS

Math League Press, P.O. Box 17, Tenafly, New Jersey 07670-0017

Contest Number 1   *Any calculator without a QWERTY keyboard is allowed.* Answers must be exact *or* have 4 (or more) significant digits, correctly rounded.   **October 15, 2019**

Name _____ Teacher _____ Grade Level _____ Score ____

Time Limit: 30 minutes   *Answer Column*

1-1.	What is the larger of two prime numbers with a sum of 2019?	1-1.
1-2.	If a 3×4 rectangle is split into eight congruent triangles as shown, what is the perimeter of one of these eight triangles?	1-2.
1-3.	If a, b, and c are 1-digit non-negative integers, not necessarily distinct, how many different values are possible for the sum $a+b+c$?	1-3.
1-4.	What is the area of an isosceles $\triangle$ with side-lengths of 22 and 61?	1-4.
1-5.	What is any pair of positive integers whose squares sum to 9797? [Hint: The identity $(a^2+b^2)(c^2+d^2) = (ac+bd)^2+(ad-bc)^2$ can help.]	1-5.
1-6.	We want 3 pizzas, each topped with 0, 1, 2, 3, 4, 5, or 6 of 6 available whole-pie toppings. How many different combinations of 3 such pizzas are possible if different pizzas need not have different toppings?	1-6.

HIGH SCHOOL MATHEMATICS CONTESTS

Math League Press, P.O. Box 17, Tenafly, New Jersey 07670-0017

Contest Number 2 *Any calculator without a QWERTY keyboard is allowed.* Answers must be exact *or* have 4 (or more) significant digits, correctly rounded. **November 12, 2019**

Name _____ Teacher _____ Grade Level _____ Score _____

Time Limit: 30 minutes | *Answer Column*

2-1. If $\frac{x-1}{x+1} = 2019$, what is the value of $\frac{x^2-2x+1}{x^2+2x+1}$? | 2-1.

2-2. What is the least prime number which can be written as a sum of two composite numbers? | 2-2.

2-3. If we define the *separation* between two points in the x-y plane as the length of the shortest path from one point to the other along the axes and/or along lines parallel to the axes, then there are exactly four points with integral coordinates whose separation from the origin is 1. How many points with integral coordinates have a separation from the origin of 5? | 2-3.

2-4. From a point inside an equilateral triangle, if the distances to the three sides are $2\sqrt{3}$, $4\sqrt{3}$ and $5\sqrt{3}$, what is the area of the equilateral triangle? | 2-4.

2-5. If a and c are rational, and if $x^3+cx^2-5x+a = (x-c)(x-c)(x+\frac{a}{c^2})$, what are all possible values of a? | 2-5.

2-6. What is the greatest real number x for which $\sqrt{x} + [x] - x = 2$, where $[x]$ is the greatest integer $\leq x$? [Note: For this question, the instructions have been changed: **your answer must be exact**.] | 2-6.

Solutions on Page 53 • Answers on Page 67

HIGH SCHOOL MATHEMATICS CONTESTS

Math League Press, P.O. Box 17, Tenafly, New Jersey 07670-0017

Contest Number 3 *Any calculator without a QWERTY keyboard is allowed.* Answers must be exact *or* have 4 (or more) significant digits, correctly rounded. **December 10, 2019**

Name _____ Teacher _____ Grade Level _____ Score ____

Time Limit: 30 minutes | *Answer Column*

3-1. For what value of x does $(x-2019)^2 = (x-2020)^2$? | 3-1.

3-2. The lengths of the sides of hexagon H are 1, 2, 3, 4, 5, and 6. If no two consecutive sides of H have consecutive-integer lengths, what is the maximum sum of the lengths of three consecutive sides of H? | 3-2.

3-3. What is the least $k \geq 2$ for which there exist k consecutive integers whose sum is 1000? | 3-3.

3-4. If $f(2t+7) = 12t+37$ for all real numbers t, what are all values of x which satisfy $f(x) = x^2$? | 3-4.

3-5. Three congruent circles have their centers on the same diagonal of a square, with two of the circles each tangent to two sides of the square, and the third circle externally tangent to the other two circles, all as shown. If the length of a side of the square is 8, what is the length of a radius of one of the circles? | 3-5.

3-6. In how many ways can Grandpa give $100 in $1 bills to his 5 grandchildren so each grandchild gets at least $10? [Note: What matters is how much each grandchild gets, not when any grandchild receives money, whether before or after another grandchild.] | 3-6.

HIGH SCHOOL MATHEMATICS CONTESTS

Math League Press, P.O. Box 17, Tenafly, New Jersey 07670-0017

Contest Number 4   *Any calculator without a QWERTY keyboard is allowed.* Answers must be exact *or* have 4 (or more) significant digits, correctly rounded.   **January 7, 2020**

Name _____ Teacher _____ Grade Level _____ Score _____

Time Limit: 30 minutes   *Answer Column*

4-1.   What is the only value of x for which $\dfrac{x^2 - 2019^2}{x - 2019} = 2020$?

4-1.

4-2.   Two of the numbers I wrote on my paper are the greatest prime numbers less than 100 that differ by 4. What is their sum?

4-2.

4-3.   For what integer k will $2^k - 1$ be the greatest divisor of $2^{22} - 2$ that is less than $2^{22} - 2$?

4-3.

4-4.   The area of the parallelogram shown is 44. If the total area of the shaded regions is 14, what is the area of the region common to the two large unshaded triangles that share a common base?

4-4.

4-5.   What is the only value of x for which there are 24 positive integers (not necessarily distinct) whose sum is x and whose product is x?

4-5.

4-6.   Point P is on the angle bisector of a base angle of an isosceles triangle whose base-length is 12 and whose leg-lengths are 10. If the distance from P to the base is 2, what is the sum of the squares of the distances from P to the three vertices of the triangle?

4-6.

HIGH SCHOOL MATHEMATICS CONTESTS

Math League Press, P.O. Box 17, Tenafly, New Jersey 07670-0017

Contest Number 5 *Any calculator without a QWERTY keyboard is allowed.* Answers must be exact *or* have 4 (or more) significant digits, correctly rounded. **February 11, 2020**

Name _____ Teacher _____ Grade Level _____ Score _____

Time Limit: 30 minutes

Answer Column

5-1. What is the sum of the reciprocals of the solutions of the equation $(x+1)(x+2)(x+3) = 0$?

5-1.

5-2. What is the largest integer n for which $2020-n$, 2020, and $2020+n$ could be the lengths of the sides of a triangle?

5-2.

5-3. The only possible scores on an exam are the 16 integers from 0 to 15. The most frequent score earned by the 100 students who took the exam was 0 (a score achieved by k students). If no other score was earned as frequently, what is the least possible value of k?

5-3.

5-4. Two isosceles triangles with supplementary vertex angles share a common base. The lengths of the legs of one triangle are 12 and of the other triangle are 5. What is the sum of the lengths of the altitudes that can be drawn to the common base of the triangles?

5-4.

5-5. Two numbers are called reversal numbers if one is obtained from the other by reversing the order of digits-for example, 123 and 321. What are the two reversal numbers whose product is 92,565?

5-5.

5-6. What are all ordered triples of non-zero integers (a,b,c) for which a, b, and c form a geometric sequence whose common ratio is a non-zero integer, while $a+4$, b, and c form an arithmetic sequence?

5-6.

© 2020 by Mathematics Leagues Inc.

Solutions on Page 56 • Answers on Page 67

HIGH SCHOOL MATHEMATICS CONTESTS

Math League Press, P.O. Box 17, Tenafly, New Jersey 07670-0017

Contest Number 6 *Any calculator without a QWERTY keyboard is allowed.* Answers must be exact *or* have 4 (or more) significant digits, correctly rounded. **March 17, 2020**

Name _____ Teacher _____ Grade Level _____ Score _____

Time Limit: 30 minutes

		Answer Column
6-1.	What is the greatest five-digit number that is divisible by 11?	6-1.
6-2.	What is the least integer $n > 0$ for which $2020 - n$ is the square of an integer?	6-2.
6-3.	In tossing a fair coin, what is the probability that the second "heads" occurs on the ninth toss?	6-3.
6-4.	The square at the right has an area of 4. Quarter-circles, centered at two opposite vertices of the square, overlap in the shaded region as shown. What is the area of the shaded region?	6-4.
6-5.	A drawing of the parabola $y = x^2$ is photographed using a microscope that magnifies by a factor of 200 in each direction. The photographed parabola is traced on (unmagnified) graph paper so the parabola's vertex, orientation, and axis of symmetry are unchanged. When compared to the drawing of $y = x^2$ (also drawn on unmagnified graph paper), for what real number a does the photograph of the parabola look like the graph of $y = ax^2$?	6-5.
6-6.	The lengths of the sides of an equilateral triangle are $\log_4 a$, $\log_{10} b$, and $\log_{25}(a+b)$, where a and b are positive numbers. What is the value of $\frac{a}{b}$?	6-6.

HIGH SCHOOL MATHEMATICS CONTESTS

Math League Press, P.O. Box 17, Tenafly, New Jersey 07670-0017

Contest Number 1 *Any calculator without a QWERTY keyboard is allowed.* Answers must be exact *or* have 4 (or more) significant digits, correctly rounded. **October 13, 2020**

Name _____ Teacher _____ Grade Level ____ Score ____

Time Limit: 30 minutes   *Answer Column*

1-1. I wrote a real number which, when divided by itself, is 2020 times the number I wrote. What number did I write?	1-1.
1-2. For how many different positive integers n is each of n, $n + 2$, and $n + 4$ a prime number?	1-2.
1-3. An 8×27 rectangle is split into four triangles, as shown at the right, by three line segments which divide the rectangle's longer sides into segments of lengths 6 and 21. How long is the dotted segment?	1-3.

21   6
8
6   21

1-4. At a company, ten employees and ten interns line up to visit the CEO in ten randomly selected pairs. If each pair of employees receives a copper ring, each pair of interns receives a brass ring, and each employee-intern pair receives a silver ring, what is the probability that the number of copper rings received equals the number of brass rings received?	1-4.
1-5. What's the only positive integer whose two largest divisors have a sum of 111?	1-5.
1-6. For how many different pairs of positive integers (a,b), with greatest common factor 1, and with $a > b$, does $ab = 30!$? [NOTE: 30! is the product of the first 30 positive integers.]	1-6.

Contest Number 2 *Any calculator without a QWERTY keyboard is allowed.* Answers must be exact *or* have 4 (or more) significant digits, correctly rounded. **November 10, 2020**

Name _____ Teacher _____ Grade Level _____ Score _____

Time Limit: 30 minutes | *Answer Column*

2-1. What is the smallest perfect square that can be written as the sum of three different prime numbers? | 2-1.

2-2. Gerry arrived at the bus stop x hours past noon. Dale arrived 4 hours later. Pat arrived at 5 P.M., x hours after Dale, all on the same date. At what time did Gerry arrive at the bus stop? [Your answer must include an A.M. or a P.M.] | 2-2.

2-3. For what value of $x > 0$ does $\dfrac{x^2 + 2021x + 2020}{x^2 - 2020x - 2021} = 2$? | 2-3.

2-4. What is the greatest integer that always divides the difference of the squares of any two different positive odd integers? | 2-4.

2-5. Of the positive integers between 1000 and 10 000 that are divisible by 8, how many have a hundreds digit of 5? | 2-5.

2-6. A square is split into four triangles, and then three of the four triangles are shaded, as shown. If the areas of the shaded triangles are 3, 4, and 6, as shown, what is the area of the unshaded triangle? | 2-6.

© 2020 by Mathematics Leagues Inc.

HIGH SCHOOL MATHEMATICS CONTESTS

Math League Press, P.O. Box 17, Tenafly, New Jersey 07670-0017

Contest Number 3 *Any calculator without a QWERTY keyboard is allowed. Answers must be exact or have 4 (or more) significant digits, correctly rounded.* **December 8, 2020**

Name _____ Teacher _____ Grade Level ____ Score ____

Time Limit: 30 minutes

		Answer Column
3-1.	When written as either 8-15-17 or 15-8-17, the date August 15, 2017 has the property that the sum of the squares of the first two numbers is equal to the square of the third number. What is the first date after August 15, 2017 that has this property?	3-1.
3-2.	If $\dfrac{46}{35} = a + \dfrac{1}{b + \dfrac{1}{c + \dfrac{1}{d}}}$, where a, b, c, and d are positive integers, what is the value of d?	3-2.
3-3.	If an integer reads the same forwards or backwards, like 404 and 1221, then it's called a *palindrome*. The first palindrome greater than 9999 is 10 001. What is the fifth palindrome greater than 9999?	3-3.
3-4.	What are all positive real numbers x which satisfy $x^{\sqrt[3]{x}} = x(\sqrt[3]{x})$?	3-4.
3-5.	Two squares are inscribed in an isosceles right triangle in different ways, as shown in the diagram. If the larger square's area is 576, what is the smaller square's area?	3-5.
3-6.	A polynomial P has non-negative integer coefficients. If $P(1) = 8$ and $P(10) = 2312$, what is $P(2)$?	3-6.

HIGH SCHOOL MATHEMATICS CONTESTS

Math League Press, P.O. Box 17, Tenafly, New Jersey 07670-0017

Contest Number 4   *Any calculator without a QWERTY keyboard is allowed.* Answers must be exact *or* have 4 (or more) significant digits, correctly rounded.   **January 5, 2021**

Name _____ Teacher _____ Grade Level _____ Score ____

Time Limit: 30 minutes   *Answer Column*

4-1.   What is the smallest composite number which is the sum of two different prime numbers?

4-1.

4-2.   If the lengths of the sides of right triangle T are 3^2, 4^2, and y, what are both possible values of y?

4-2.

4-3.   What is the perimeter of a square whose vertices, as shown, are midpoints of alternating sides of a regular octagon whose perimeter is $16\sqrt{2}$?

4-3.

4-4.   Of the positive integers less than 2021, how many can be written as a difference of two powers of 2?

4-4.

4-5.   What is the least number n with the property that, in every group of n people, there are at least three people who are all friends (each knows the other two) or all strangers (none of them knows either of the other two)?

4-5.

STEVE'S DINER

4-6.   The numbers 12, 34, 56, 78, 90 are five two-digit numbers that use all ten digits. Which five two-digit numbers that use all ten digits have the largest possible product?

4-6.

HIGH SCHOOL MATHEMATICS CONTESTS

Math League Press, P.O. Box 17, Tenafly, New Jersey 07670-0017

Contest Number 5   *Any calculator without a QWERTY keyboard is allowed.* Answers must be exact *or* have 4 (or more) significant digits, correctly rounded.   **February 9, 2021**

Name _____ Teacher _____ Grade Level _____ Score _____

Time Limit: 30 minutes   *Answer Column*

5-1. Pat found the average of the first 2021 positive even integers, and Lee found the average of the first 2021 positive odd integers. How much greater is Pat's average than Lee's?

5-1.

5-2. When $N = 10^{100} - 10^{50} - 1$ is expressed as an integer in standard form, what is the sum of the digits of N?

5-2.

5-3. What is the greatest possible length of a side of a triangle whose perimeter is 1000 and all of whose sides have integral lengths?

5-3.

5-4. If i is the imaginary unit, what are all ordered pairs of integers (a,b) for which $a+bi$ is a solution of $x^2+3x+3 = i$?

5-4.

5-5. Isosceles $\triangle T$ has legs of length 120. Points on the legs split each leg into segments whose lengths are shown in the accompanying diagram (which is not drawn to scale). The little triangle at the bottom, the one which has T's base as one of its sides, has an area of 17. What is the area of $\triangle T$?

45 64
45 32
30 24

5-5.

5-6. Each round, I flip three fair coins and you flip two fair coins. Each round, we might throw equal numbers of heads. In the round that we first flip different numbers of heads, what is the probability that I threw more heads?

5-6.

Solutions on Page 62 • Answers on Page 67

HIGH SCHOOL MATHEMATICS CONTESTS

Math League Press, P.O. Box 17, Tenafly, New Jersey 07670-0017

Contest Number 6 *Any calculator without a QWERTY keyboard is allowed.* Answers must be exact *or* have 4 (or more) significant digits, correctly rounded. **March 16, 2021**

Name _____ Teacher _____ Grade Level _____ Score _____

Time Limit: 30 minutes | *Answer Column*

6-1. If x is real, what is the greatest possible value of $\dfrac{4042}{2021x^{2020} + 2}$? | 6-1.

6-2. Two coplanar congruent regular 15-gons share a side in common, but have no interior points in common, as shown. What is the degree-measure of the angle marked x? | 6-2.

6-3. If $\log_{10}(1) + \log_{10}(2) + \log_{10}(3) + \ldots + \log_{10}(2000) = 5735.52\ldots$, then how many digits are there in the expansion of 2000!, where the exclamation point represents "factorial"? | 6-3.

6-4. If the roots of $x^3 + ax^2 + bx + c$ are 1, 2, and 3, what are the roots of
$$(x-2)^3 + a(x-2)^2 + b(x-2) + c = 0?$$ | 6-4.

6-5. My three cars can travel the same distance on a tank of gas. My second car gets 6 kilometers per liter more than my first, but its tank holds 3 fewer liters than my first. My third car gets 6 kilometers per liter less than my first, but its tank holds 6 more liters than my first. How many kilometers can each car travel on a tank of gas? | 6-5.

6-6. By drawing a line parallel to its bases, I can split trapezoid T into two new trapezoids whose areas are equal. If the bases of T have lengths 2 and 14, how long is the segment common to the new trapezoids? | 6-6.

© 2021 by Mathematics Leagues Inc.

Complete Solutions

October, 2016 – March, 2021

Problem 1-1

Since any integer that divides each of two integers also divides their difference, the greatest common divisor of 2016 and 2017 divides $2017 - 2016 = 1$, whose greatest divisor is $\boxed{1}$.

Problem 1-2

By substitution, we can determine that the sum of the digits, $1 + A + 7$, is a multiple of 3 but not a multiple of 9 for $A = \boxed{4, 7}$.

Problem 1-3

In each small rectangle, if the length is ℓ and the width is w, then the perimeter is $2(\ell + w) = 18$. The length of each side of the large square is $\ell + w = 9$, so the area of the large square is $(\ell + w)^2 = 9^2 = \boxed{8}$.

Problem 1-4

If the length of one of the three sides is the average of the lengths of the other two sides, the three sides from a three-term arithmetic sequence. It's easy to show (see below) that if a right triangle's three side-lengths form an arithmetic sequence, those lengths are in the ratio 3:4:5. If the side-lengths are $3x$, $4x$, and $5x$, the perimeter is $12x$. Divide each length by $12x$ to get a triangle with perimeter 1 and side-lengths $3x/12x = 1/4$, $4x/12x = 1/3$, and $5x/12x = 5/12$. The area of the triangle is $(1/2)(1/4)(1/3) = \boxed{1/2}$.

[**PROOF:** Let the length of the longer leg be a, and let the other two side-lengths be $a-d$ and $a+d$. By the Pythagorean Theorem, $(a+d)^2 = a^2 + (a-d)^2$. Collecting terms and solving for a, we get $a = 4d$. The lengths of the other sides are $a-d = 3d$ and $a+d = 5d$, so the side-lengths are $3d$, $4d$, and $5d$.]

Problem 1-5

To maximize the number of guests, minimize the number of almonds each ate (under the condition that no more than 2 guests ate the same number of almonds), with 2 guests eating 1 almond, 2 eating 2 almonds, 2 eating 3 almonds, etc. As a consequence, $2(1+2+3+...+n) \le 2600$, or $1+2+...+n \le 1300$. Summing on the left and then clearing denominators, $n(n+1) \le 2600$. If $n = 50$, the sum is 2550. Even a single additional person would bring the sum to at least 51 more, which exceeds 2600, so the greatest possible value of n is 50. Since 2 guests ate each number of almonds, at most $\boxed{100}$ guests are needed to consume 2550 almonds.

Problem 1-6

Let a_n represent the value of the nth term. Since n is an integer from 1 through 11, and since each $a_n \ge 12$ must be a multiple of n, it follows that, when $n \ge 2$ every a_n is composite. Starting with 12, the ten smallest composite numbers (the values of a_n) are 12, 14, 15, 16, 18, 20, 21, 22, 24, and 25. Can we pair these ten integers (the a_n values) with their term numbers (the subscripts greater than 1)? Among these ten, the only multiples of 9, 10, and 11 are 18, 20, and 22 respectively, so $18 = a_9$, $20 = a_{10}$, and $22 = a_{11}$. The only factor of 25 in the interval $2 \le n \le 11$ is 5, so $25 = a_5$. That forces $15 = a_3$, which in turn mandates $21 = a_7$ and then $14 = a_2$. Three terms remain. There are several possibilities. For example, if $a_8 = 16$, then a_4 and a_6 get assigned to 12 and 24 in either order. What about the first term, a_1? All integers are multiples of 1, so the eleven-term sequence will have its minimum sum when the value of a_1 is minimized. The least integer greater than 12 that is not yet in the sequence is 13. Finally, the sum of all eleven terms is at least $12 + 13 + 14 + 15 + 16 + 18 + 20 + 21 + 22 + 24 + 25 = \boxed{200}$.

Contests written and compiled by Steven R. Conrad, Daniel Flegler, & Adam Raichel   ©2016 by Mathematics Leagues Inc.

Problem 2-1

From the statement of the problem, we are told that $d + 3 = (d - 3)^2$, so $d^2 - 7d + 6 = (d-6)(d-1) = 0$. The solutions to this equation are $d = 1$ and $d = 6$. Since $d = 1$ must be rejected, we see that $d = \boxed{6}$.

Problem 2-2

The 3-4-5 right triangle is the only right triangle with integral side-lengths whose hypotenuse has length 5. If the rectangle's width were 4, its length would be $5+3 = 8$ and its area would be $4 \times 8 = 32$. If the rectangle's width were 3, its length would be $5+4 = 9$ and its area would be $\boxed{27}$.

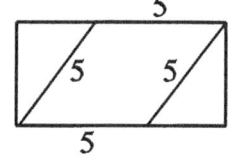

Problem 2-3

Adding $1159x + 857y = 2798$
and $\underline{857x + 1159y = 1234}$, we get
$2016x + 2016y = 4032$, so $x+y = \boxed{2}$.

Problem 2-4

Since $k > 0$ and k is even power of an even number, k must be a multiple of 4. Therefore, we can let $3^k = 3^{4m} = (3^4)^m = 81^m$, whose units digit is always $\boxed{1}$.

Problem 2-5

We know that $f(x) = 3f(1-x)+1$ for all x. Let $t = 1-x$, from which $x = 1-t$. Substituting into the original equation, $f(1-t) = 3f(t)+1$, from which $f(1-x) = 3f(x)+1$. Substitute this into the first equation to get $f(x) = 3[3f(x)+1]+1 = 9f(x)+3+1$, or $8f(x) = -4$. Solving, $f(x) = \boxed{-\frac{1}{2}}$ for every value of x.

Problem 2-6

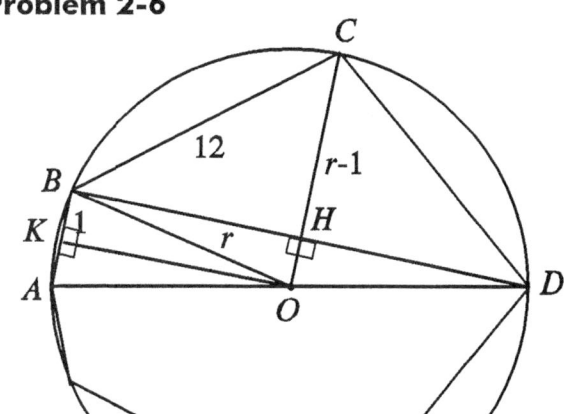

Draw diameter $\overline{AD}$, radii $\overline{OB}$ and $\overline{OC}$, and chord $\overline{BD}$. Let the perpendicular drawn from O to $\overline{AB}$ intersect $\overline{AB}$ at K. Since $\overline{BC} \cong \overline{CD}$ and $\overline{OB} \cong \overline{OD}$, points O and C are equidistant from points B and D. Let H be the point at which $\overline{OC}$ is perpendicular to and bisects $\overline{BD}$. Also, since $\angle ABD$ is inscribed in a semicircle, $\angle ABD$ is a right angle. Since quadrilateral $KOHB$ has 3 right angles, it must be a rectangle. Since the altitude to the base of an isosceles triangle bisects the base, $KB = 1 = OH$. Also, $HC = OC - OH = r-1$. In $\triangle OHB$, $BH^2 = OB^2 - OH^2 = r^2 - 1$. By the Pythagorean Theorem, $BH^2 + HC^2 = BC^2 = 12^2$, so $(r^2-1)+(r-1)^2 = 144$. Expanding and solving, $r^2 - r - 72 = (r-9)(r+8) = 0$, so $r = 9$ and the area of the circle is $\boxed{81\pi}$.

[**NOTE:** Alternatively, it is possible to use the law of cosines to achieve the same result.]

Problem 3-1

Set up the division algorithm and begin dividing. The quotient starts off as 1 04 1 0? The final 2-digit number will be divisible by 7 if the final 2 digits are 14. so the missing digit is a $\boxed{4}$.

Problem 3-2

Since January has 31 days, there will be four complete cycles of the seven-day week plus three additional days. The consecutive days of the week upon which these three additional days fall will be occurring for the fifth time in the month. If these three consecutive days do not include Monday or Friday, they must be Tuesday, Wednesday, and Thursday. January 1 fell on a $\boxed{\text{Tuesday}}$.

Problem 3-3

All multiples of 15 are multiples of 3, as are all multiples of 21. Every circled number will be a multiple of 3, and the least possible difference between two distinct multiples of 3 is $\boxed{3}$.

[We want to minimize $M = |15x - 21y| > 0$, where x and y are positive integers. Since $M = 3|5x - 7y|$ is a positive multiple of 3, the minimum is $3 \times 1 = 3$. This minimum will occur whenever $|5x - 7y| = 1$. Two examples are $(3,2)$ and $(4,3)$.]

Problem 3-4

"If n is divided by 2016, the remainder is 1008" translates into $\frac{n}{2016} = t + \frac{1008}{2016}$, where t is an integer. If we divide $2n$ by 2016, we get $\frac{2n}{2016} = 2t + \frac{2016}{2016} = 2t+1$, an integral quotient with a remainder of $\boxed{0}$.

Problem 3-5

If a line parallel to the width of the horizontal rectangle is drawn to create a right triangle with hypotenuse equal to the side of the shaded parallelogram, the sides of the triangle formed must be 8, 8, and $8\sqrt{2}$. Using the $8\sqrt{2}$ side of the parallelogram as its base, the parallelogram's height would be 6 and its area would be $\boxed{48\sqrt{2}}$.

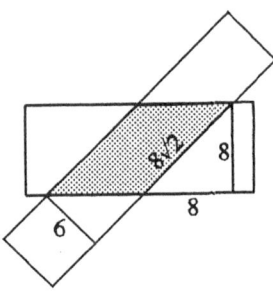

Problem 3-6

By symmetry, a circle with center $(\sqrt{2}, \sqrt{2})$ and a radius just long enough for C to pass through $(1,0)$ must also pass through $(0,1)$. Can there be 3 such points? If so, there would be a triangle T inscribed in the circle whose vertices all had rational (x,y) coordinates. That would mean all non-vertical sides of T would have rational slopes and also that the midpoint of each side of T would have rational coordinates. The perpendicular bisectors of T's non-horizontal sides would have rational slopes, so they will intersect at a point with rational coordinates. That means that the center of T's circumscribed circle has rational coordinates. Since the center of C has irrational coordinates, we have proved that the largest value of n is $\boxed{2}$.

Another way to say this is that any 2 distinct rational points determine a rational line (a line of the form $ax + by + c = 0$, with a,b,c rational), while any 2 non-parallel rational lines intersect at a rational point.

Contests written and compiled by Steven R. Conrad, Daniel Flegler, & Adam Raichel   ©2016 by Mathematics Leagues Inc.

Problem 4-1

Each of the 2017 terms of $2 + 4 + 6 + \ldots$ is 1 more than the corresponding term of $1 + 3 + 5 + \ldots$, so the sum of the first 2017 even integers exceeds that of the first 2017 odd integers by $\boxed{2017}$.

Problem 4-2

By trial and error, 3 does not work since $1+3 = 4$. Continuing, the sum of the divisors of 4 is $1 + 2 + 4 = 7$, so the least positive integer greater than 2 whose divisor-sum is a prime is $\boxed{4}$.

[NOTE: Positive integers with a prime divisor-sum must be expressible as a power of a prime. Examples include $3^2 = 9$, $2^4 = 16$, and $5^2 = 25$, but **not** 7^2.]

Problem 4-3

The probability of getting

A) no bullseye in 1 try $= \frac{1}{2}$

B) at least 1 bullseye in 2 tries $= \frac{1}{2} + \frac{1}{2} \times \frac{1}{2} = \frac{3}{4}$

C) at least 2 bullseyes in 3 tries $= 3 \times \frac{1}{8} + \frac{1}{8} = \frac{1}{2}$.

The most likely of these options is $\boxed{B}$.

Problem 4-4

The larger (and other) unshaded circles have respective radius-lengths 3 and r. The shaded region's area = the area of circle C − the sum of the areas of the unshaded circles $= \pi(3+ r)^2 - (9\pi + \pi r^2) = 6\pi r$. When 2 chords intersect inside a circle, the product of the segment-lengths of one = the product of the segment-lengths of the other. Consequently, $6 \times 2r = 4 \times 4$, $r = 4/3$, and $6\pi r = \boxed{8\pi}$.

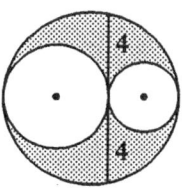

[NOTE: It's not too difficult to prove the surprising fact that the area of the shaded region is independent of the radius-lengths of the unshaded circles, as long as the product of their radius-lengths is 4.]

Problem 4-5

Method I: The equation is a quadratic in $(\log_{10}x)$ whose discriminant $= 36+4a \geq 0$ for $(\log_{10}x)$ to be a real number. The least such real a is $\boxed{-9}$.

Method II: Complete the square to get $(\log_{10}x)^2 + 6\log_{10}x + 9 = a + 9$, so $(\log_{10}x + 3)^2 = a + 9$. Since $\log_{10}x$ can be any real number, the left side is the square of a real number, so the left side is ≥ 0. Therefore, the left side = the right side $= a+9 \geq 0$. The least such real a is -9.

Problem 4-6

There are 1999 terms, and the term in the middle is

$$\frac{4^{\frac{1000}{2000}}}{4^{\frac{1000}{2000}} + 2} = \frac{2}{2+2} = 0.5.$$ By pairing the nth and the

$(2000-n)^{\text{th}}$ terms, the sum of the other 1998 terms is

$$\left(\frac{4^{\frac{1}{2000}}}{4^{\frac{1}{2000}} + 2} + \frac{4^{\frac{1999}{2000}}}{4^{\frac{1999}{2000}} + 2} \right) + \left(\frac{4^{\frac{2}{2000}}}{4^{\frac{2}{2000}} + 2} + \frac{4^{\frac{1998}{2000}}}{4^{\frac{1998}{2000}} + 2} \right) +$$

$$\ldots + \left(\frac{4^{\frac{999}{2000}}}{4^{\frac{999}{2000}} + 2} + \frac{4^{\frac{1001}{2000}}}{4^{\frac{1001}{2000}} + 2} \right).$$ As long as no denominator is 0, $\frac{a^k}{a^k + 2} + \frac{a^{1-k}}{a^{1-k} + 2}$ is always 1, so the sum

of each pair's terms is 1, so $S = 999+0.5 = \boxed{999.5}$.

Contests written and compiled by Steven R. Conrad, Daniel Flegler, & Adam Raichel   ©2017 by Mathematics Leagues Inc.

Problem 5-1

Since the hypotenuse must be the longest side, if hypotenuse$^2 = 20$, then the leg $= \sqrt{20-16} = 2$. If leg$^2 = 20$, then the hypotenuse $= \sqrt{20+16} = 6$. The possible lengths of the third side are $\boxed{2, 6}$.

Problem 5-2

Excluding 2, 3, 5, 7, 11, 13, and 17, the only prime whose square is less than 500 is 19. Therefore, the only perfect square between 2 and 500 which is not divisible by any prime less than 19 is $19^2 = \boxed{361}$.

Problem 5-3

We're asked about $\frac{1}{b} + a$. We're told that $\frac{b}{1+ab} = 2017$. To get b in the denominator, take the reciprocal of both sides to get $\frac{1+ab}{b} = \frac{1}{b} + a = \boxed{\frac{1}{2017}}$.

Problem 5-4

Dividing, $z^{20}/z^{12} = z^8 = 1$. Similarly, $z^{12}/z^8 = z^4 = 1$. If $z^4 = 1$, then $z = \boxed{\pm 1, \pm i}$. All steps are reversible.

[Note that $z^{20} = (z^4)^5$ and $z^{12} = (z^4)^3$. Therefore, if $z^4 = 1$, then $z^{12} = 1$ and $z^{20} = 1$.]

Problem 5-5

The two small right triangles, one shaded and one unshaded, have two congruent angles, so the

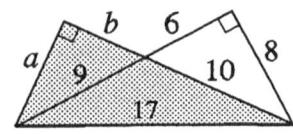

triangles are similar. Since the sides of similar triangles are proportional, $\frac{9}{10} = \frac{b}{6} = \frac{a}{8}$. Solving, $a = 7.2$ and $b = 5.4$. The area of the shaded large right triangle $= \frac{1}{2}(a)(b+10) = \frac{1}{2}(7.2)(15.4) = \boxed{55.44}$.

Problem 5-6

Since $\frac{1}{m} + \frac{1}{n} + \frac{1}{mn} = \frac{1}{5}$, clear fractions to get $mn = 5m+5n+5$. Rearrange the terms as if trying to factor to get $mn-5m-5n = 5$. The left side is almost the expansion of $(m-5)(n-5)$; so add 25 to both sides to get $(m-5)(n-5) = 30$. Since $m-5$ and $n-5$ are positive factors of 30, $m-5$ can be 1, 2, 3, 5, 6, 10, 15, or 30. Each value of m generates a value of n, so the total number of ordered pair solutions is $\boxed{8}$.

Contests written and compiled by Steven R. Conrad, Daniel Flegler, & Adam Raichel   ©2017 by Mathematics Leagues Inc.

Problem 6-1

$(x-2018)(2016-x) = (2016-x)(-1)(-1)(x-2018) = (x-2016)(2018-x) = \boxed{1}$.

Problem 6-2

In the Venn diagram, the unshaded portion represents Me crying and You not crying. The shaded portion represents You crying. 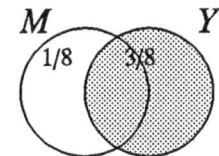 Half the time, at least one of them cries on a Friday. To determine the probability that You cries on a Friday, solve $\frac{1}{2} = \frac{1}{8} + Y$ to get $Y = \boxed{\frac{3}{8}}$.

Problem 6-3

By observation, one solution is (0,0). If neither a nor b is 0, then divide by ab to get $\frac{a}{b} + \frac{b}{a} = 1$. Since the two fractions have the same sign and their sum is 1, a positive number, a and b are both positive or both negative. If $a = b$, the sum is 2. If $a \neq b$, one of the fractions is greater than 1, and the other is between 0 and 1. In fact, the sum of any non-zero real number and its reciprocal is at least 2 in absolute value, so $\frac{a}{b} + \frac{b}{a} = 1$ has no real solutions, and the total number of solutions is $\boxed{1}$.

Problem 6-4

Of the twelve 30°-60°-90° triangles shown in the diagram, 4 are shaded. All 12 triangles are congruent, so the ratio of the area of the shaded region to that of the hexagon is $\boxed{\frac{1}{3}}$.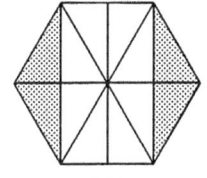

Problem 6-5

Let the longer segment of the longer leg have length x, so the length of the longer leg is $x+1$. The bisector of an angle of a triangle divides the opposite side into segments proportional to the adjacent sides, so $\frac{h}{x} = \frac{2}{1}$. Substitute $h = \sqrt{2^2 + (1+x)^2}$, multiply by x, square, and collect terms to get that $3x^2 - 2x - 5 = (3x-5)(x+1) = 0$, so $x = \frac{5}{3}$ and the length of the longer leg $= x + 1 = \boxed{\frac{8}{3}}$.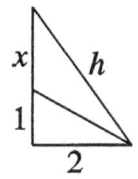

Problem 6-6

Let $d > 0$ be the constant difference between consecutive terms of the sequence whose first term is a. The sequence described in the question is the arithmetic progression $\{a + dk : k = 0, 1, \ldots, 49\}$. If the first term a were less than 50, then $a + dk$ would become $a + da = a(1+d)$ when $k = a$, and this is divisible by a. Since $a > 1$, we contradict the assumption that the first term is relatively prime to the other terms. Therefore $a \geq 50$. If a is even, there are many even terms. This is not possible. Let's try odd number values for a.

If $a = 51$ then in the sequence $51, 51+d, 51+2d, 51+3d$, the terms 51 and $51+3d$ are both divisible by 3, so this is not possible.

If $a = 53$, a prime, then the resulting sequence is $53, 53+d, 53+2d, \ldots, 53+49d$. To minimize the last term we try $d = 1$, and the sequence becomes $53, 54, 55, \ldots, 102$. Since $53 \times 2 = 106 > 102$, every term after 53 is relatively prime to 53, as required, so the least possible value of $53 + 49d$ is $53 + 49 = \boxed{102}$.

Contests written and compiled by Steven R. Conrad, Daniel Flegler, & Adam Raichel   ©2017 by Mathematics Leagues Inc.

Problem 1-1

Since $\dfrac{1}{x+2017} = 1$, $x+2017 = 1$, so $x+2018 = 2$.

Now take reciprocals to get $\dfrac{1}{x+2018} = \boxed{\dfrac{1}{2}}$ or 0.5.

Problem 1-2

If the number of pirates is P and the number of treasure chests is T, then $P = T+1$ and $P/2 = T-1$, so $T = \boxed{3}$.

Problem 1-3

If the lengths of the sides of the rectangle are the two largest primes less than 120, then the perimeter of the rectangle is $2(109 + 113) = \boxed{444}$.

Problem 1-4

The small square that shares borders with the two squares at the top must have sides of length $18 - 14 = 4$. On the left side, the square in the middle has sides of length $14 - 4 = 10$. To the right of the 10×10 square and the 1×1 square are the 4×4 square and a 7×7 square. The width of the rectangle is $14 + 18 = 32 = 10+7+x$, so $x = 15$ and the area of the shaded region is $\boxed{225}$.

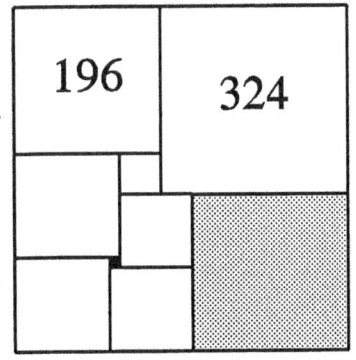

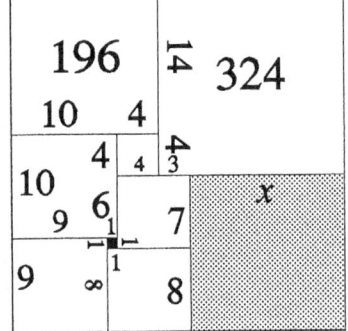

Problem 1-5

For every quadruple of distinct points, there are 3 ways to partition them into pairs, 2 pairs of which do not intersect. Therefore, the probability is $\boxed{\dfrac{2}{3}}$.

(Draw a convex quadrilateral and its diagonals to see that the 4 vertices determine 3 pairs of line segments, 2 of which do *not* intersect.)

Problem 1-6

We'll show that there's a value of N beyond which one can always find a solution by using a previous solution. For example, if (a,b) is a pair of **non-negative** integers that satisfies $7x+11y = N$ and if $a \geq 3$, then $(a-3,b+2)$ satisfies $7x+11y = N+1$, because replacing three 7s with two 11s increases N by 1. And if $b \geq 5$, then $(a+8,b-5)$ satisfies $7x+11y = N+1$, because replacing five 11s with eight 7s increases N by 1. This leaves unanswered the question of what happens with a solution like $(2,4)$ in which both a and b are too small to carry out the above procedure. At $(2,4)$, $7x+11y = 58$, and we have no procedure to increase N by 1 to 59. That's because $7x+11y = 59$ has no solution with x and y non-negative integers. (Any such solution requires that $y \leq 5$. By trial and error, no such solution exists.) On the other hand, for the next 7 consecutive values of N (60, 61, 62, 63, 64, 65, and 66), the solutions of $7x+11y = N$ are $(7,1)$, $(4,3)$, $(1,5)$, $(9,0)$, $(6,2)$, $(3,4)$, and $(0,6)$. Solutions continue forever because if (a,b) is a solution of $7x+11y = N$, then $(a+1,b)$ is a solution of $7x+11 = N+7$. Since this proves that there is a solution for every $N \geq 60$, the largest **unattainable** value of N is $\boxed{59}$.

[By a theorem of *Frobenius*, if p and q are positive integers with no common integer factor > 1, and if x and y are non-negative integers, then the largest positive integer that is *not* of the form $px + qy$ is $pq - p - q$.]

Contests written and compiled by Steven R. Conrad, Daniel Flegler, & Adam Raichel   ©2017 by Mathematics Leagues Inc.

Problem 2-1

Setting each factor equal to 0 and solving, the solutions are –2017, –2018, –2019, and –2020, the largest of which is $\boxed{-2017}$.

Problem 2-2

As the values of m and n that satisfy $\sqrt{m} + \sqrt{n} = 10$ get further apart, their sum gets larger. Since the largest sum will come from minimizing one of the variables with a value of 1, while maximizing the other with a value of 81, the greatest possible value of $m+n$ is $81+1 = \boxed{82}$.

Problem 2-3

If the average of the 3 positive integers is 20, their sum is $3 \times 20 = 60$. The average of the two smallest is 8, so their sum is 16. The third integer is $60-16 = 44$. To maximize $d-m$, make m as small as possible. To do that, let the 3 numbers be 1, 15, and 44. The greatest possible value of $d-m$ is $44-1 = \boxed{43}$.

Problem 2-4

The difference between two squares is the product of their sum and their difference. The sum and difference of two integers have the same *parity* (they are both even or they are both odd). Since the digit is even, both factors must be even. The product of 2 even numbers must be divisible by 4, so the only possible even digits are $\boxed{4,\ 8}$.

[**NOTE:** $11\,111\,112^2 - 11\,111\,110^2 = 44\,444\,444$, and $11\,111\,113^2 - 11\,111\,109^2 = 88\,888\,888$.]

Problem 2-5

ABCD is a parallelogram with an area of 90, so the area of $\triangle BCD$ is 45, and $BC = AD = 3x$. Since $\triangle PCB \sim \triangle PED$, their sides

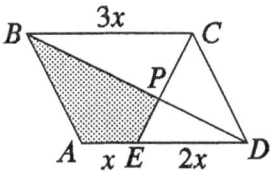

are in the ratio 3:2. Let's use this fact repeatedly. The sum of the areas of $\triangle PCB$ and $\triangle PCD$ is 45. Their bases, $\overline{BP}$ and $\overline{DP}$, are in the ratio 3:2. In $\triangle BCP$ and $\triangle DCP$, the altitudes to $\overline{BD}$ are the same, so the ratio of the areas of the two triangles equals the ratio of their base-lengths = 3:2. The area of $\triangle PCB$ is $(3/5)(45) = 27$, and the area of $\triangle PCD$ is 18. Since $PC:PE = 3:2$, the areas of $\triangle PCD$ and $\triangle PED$ are in the ratio 3:2, so the area of $\triangle PED$ is 12. To get the shaded region's area, subtract the area of $\triangle PED$ from the area of $\triangle ABD$ to get $45-12 = \boxed{33}$.

Problem 2-6

We seek such n for which n^2-n is divisible by 10^6. We will use the convenient fact that $n^2-n = -n(1-n)$ is unchanged if we replace n with $1-n$: for $m = 1-n$, we have $n(1-n) = (1-m)m$. so negating both sides gives $n^2-n = m^2-m$. We're told that if $n = 890\,625$, then n^2-n is divisible by 10^6; so likewise for $m = 1-890\,625 = -890\,624$, we know that m^2-m is divisible by 10^6. Unfortunately, m is negative. By adding 10^6 to m, we can fix this problem without the loss of this crucial divisibility property. To prove this, we use $(x + 10^6)^2 - (x + 10^6) = (x^2 - x) + 10^6(2x + 10^6 - 1)$. Since $m+10^6 = -890\,624+1\,000\,000 = \boxed{109\,376}$ is a number in the desired range that has the divisibility property, it is the sought-after second integer.

[**NOTE (for very advanced students):** The seed of both this problem and its solution is the Chinese Remainder Theorem, which underlies the proof that there are only two such values of n.]

Contests written and compiled by Steven R. Conrad, Daniel Flegler, & Adam Raichel   ©2017 by Mathematics Leagues Inc.

Problem 3-1

Subtracting 40^2 from both sides, $a^2 = 50^2 - 40^2 = 900$, so $a = \pm 30$. The least value of a is $\boxed{-30}$.

Problem 3-2

We are told four things: 1) $y = 2x$, 2) both x and y are positive, 3) $x+y$ is the square of an integer, and 4) x is as small as possible, subject to these conditions. Therefore, $x + 2x = 1$, so $x = \frac{1}{3}$, $y = \frac{2}{3}$, and $(x,y) = \boxed{\left(\frac{1}{3}, \frac{2}{3}\right)}$.

Problem 3-3

One of the 9 different numbers chosen at random is the smallest of the 9, so the probability that it was the last number chosen is $\boxed{\frac{1}{9}}$.

Problem 3-4

Let the lengths of the sides be x, x, and $2017-2x$. Using the triangle inequality, we get $2x > 2017-2x$, so $x > 504.25$. Similarly, from $x + (2017-2x) > x$, we get $x < 1008.5$. Thus, $504.25 < x < 1008.5$, an interval containing $\boxed{504}$ integers.

Problem 3-5

Rewrite $y = ab^2 - a^2b$ as $y = -b(a^2) + b^2(a)$, a quadratic in a. When graphed in the a–y plane, the vertex V of this parabola is at the point V for which $a = \frac{-b^2}{-2b} = \frac{b}{2}$. At V, $-ba^2 + b^2a = -\frac{b^3}{4} + \frac{b^3}{2} = \frac{b^3}{4}$. In addition, since $b \le 1$, $\frac{b^3}{4} \le \boxed{\frac{1}{4}}$, which is a value that actually occurs when $b = 1$ and $a = \frac{1}{2}$.

Problem 3-6

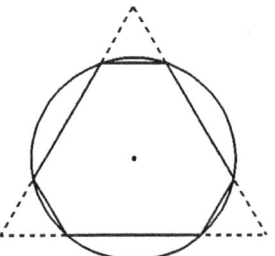

First we'll prove that the area of the hexagon is independent of the order in which its sides are drawn. Draw radii connecting the center of the circle to the hexagpn's vertices, creating 6 isosceles triangles: 3 (all congruent) with base-length 2 and 3 (all congruent) with base-length 1. Regardless of how the hexagon's sides are arranged, its area is the sum of the areas of the 6 triangles. For simplicity, draw the hexagon with sides alternating in length, as shown. Two consecutive sides of this hexagon cut off one-third of the circle, so each angle of the hexagon is 120°. Extend the hexagon's longer sides to create an equilateral triangle of side-length 4. The hexagon's area = the area of this large equilateral triangle minus the areas of the 3 smaller equilateral triangles (each of side-length 1) formed by these extensions and the hexagon's shorter sides. The required area is $16\sqrt{3}/4 - 3(\sqrt{3}/4) = \boxed{13\sqrt{3}/4} = 5.629165 \ldots$.

Contests written and compiled by Steven R. Conrad, Daniel Flegler, & Adam Raichel　©2017 by Mathematics Leagues Inc.

Problem 4-1

Since the area is 2018 times the perimeter, $s^2 = 2018(4s) = 8072s$, from which $s = \boxed{8072}$.

Problem 4-2

The shaded square is bordered by an isosceles right triangle on the right and by another on the left. If the length of a side of the shaded square is x, then $3x = 12$, $x = 4$, and $x^2 = \boxed{16}$.

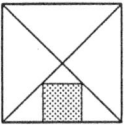

Problem 4-3

The minimal answer is $\boxed{4,5,10,20,40}$.

There happens to be a general solution to this problem. In the sequence a, $1+a$, $2+2a$, $4+4a$, $8+8a$, if the second term is divisible by 5, so are the last three terms, so a general solution is $5n-1$, $5n$, $10n$, $20n$, $40n$.

Problem 4-4

Since 1 is not a prime, $r > 2$. Since r is odd, $r-1$ is even, so $p = 2$. If $q > 1$ were odd, the expression on the left would have the form $2^{odd}+1$, which is factorable for all odd integers > 1. Therefore q must also be 2, so $(p,q,r) = \boxed{(2,2,5)}$.

Problem 4-5

If $5x^3+2xy = 23$, then $x(5x^2+2y) = 23$. Since x is an integer, x is an integral divisor of 23. Since 23 is prime, the only integral divisors of 23 are 1, –1, 23, and –23. Substitute each of these into the original equation and then solve for y to get that the only four solutions are $\boxed{(1,9),\ (-1,-14),\ (23,-1322),\ (-23,-1323)}$.

Problem 4-6

Using brackets to denote the greatest integer function,

If $1/2 < x < 1$, then $[\log_2\left(\frac{1}{x}\right)] = 0$.

If $1/4 < x < 1/2$, then $[\log_2\left(\frac{1}{x}\right)] = 1$.

If $1/8 < x < 1/4$, then $[\log_2\left(\frac{1}{x}\right)] = 2$.

If $1/16 < x < 1/8$, then $[\log_2\left(\frac{1}{x}\right)] = 3$.

$\vdots$

If $1/2^{n+1} < x < 1/2^n$, then $[\log_2\left(\frac{1}{x}\right)] = n$.

To find the sum of the lengths of the intervals in which the number on the right is odd, add together

$$1/4 + 1/16 + 1/64 + \ldots = \frac{1/4}{1 - 1/4} = \boxed{\frac{1}{3}}.$$

Contests written and compiled by Steven R. Conrad, Daniel Flegler, & Adam Raichel   ©2018 by Mathematics Leagues Inc.

Problem 5-1

The product of two numbers with a fixed sum increases as the difference between the numbers decreases. Let's start with $15 \times 15 = 225$. The two numbers are not prime. How about $14 \times 16 = 224$? Again, neither is a prime. Finally, since both 13 and 17 are prime, the answer is $13 \times 17 = \boxed{221}$.

Problem 5-2

From the common vertex of the two shaded triangles to either horizontal side of the rectangle is half the height of the rectangle. The base of each triangle opposite that common vertex is one-third the length of a horizontal side. Therefore, the two triangles together occupy $(1/2)(1/3) = 1/6$ of the rectangle's area, so the rectangle's area is $6 \times 2018 = \boxed{12108}$.

Problem 5-3

Let the respective probabilities be x, $2x$, and $4x$. The probability that it is the third beaver that takes the last bite is $\dfrac{4x}{x+2x+4x} = \boxed{\dfrac{4}{7}}$.

Problem 5-4

Let (m,n) be transformed to (a,b). The segment joining them is horizontal, so $n = b$. Also, m and a are equidistant from $x = 4$, so $|m-4| = |a-4|$. Either $m-4 = a-4$ or $m-4 = 4-a$. Solving, $m = a$ or $a = 8-m$. Since we were told that $m \neq 4$, it follows that $m \neq a$, so $a = 8-m$. Finally, the reflection of (m,n) is $(a,b) = \boxed{(8-m,n)}$.

Problem 5-5

Let the lengths of the sides of the triangle be a, b, and c, with $c > b > a$. If the length of the altitude to c is h, then h must be the shortest altitude since it is drawn to the longest side, and $h < 10$. Writing the area of the triangle 3 ways, $15a = 10b = hc > 0$. Thus, $a = \dfrac{hc}{15}$ and $b = \dfrac{hc}{10}$. By the triangle inequality, $a+b > c$. Substituting, we get $\dfrac{hc}{15} + \dfrac{hc}{10} > c$. Dividing through by c, $\dfrac{h}{15} + \dfrac{h}{10} > 1$. Solving, $h > 6$. Therefore, the least possible integer value of h is $\boxed{7}$.

Problem 5-6

Method I: $x^4-4x^3+2x^2-4x+1 = x^2(x^2-4x+1) + (x^2-4x+1) = (x^2+1)(x^2-4x+1) = 0$. For a real root, we must solve $x^2-4x+1 = 0$ to get $x = 2 \pm \sqrt{3}$. The sine function has a maximum value of 1, so $\sin\theta = 2-\sqrt{3}$, and $\cos 2\theta = 1-2\sin^2\theta = 8\sqrt{3}-13$. For integral coefficients, square $\cos 2\theta + 13 = 8\sqrt{3}$ to get $\cos^2 2\theta + 26\cos 2\theta - 23 = 0$; so $(a,b) = \boxed{(26,-23)}$.

Method II: If a and b are integers and, as above, $\cos 2\theta = x = -13+8\sqrt{3}$ is one root, the conjugate, $-13-8\sqrt{3}$ is a second root. The quadratic equation with lead coefficient 1 for which these are roots is $[x-(-13+8\sqrt{3})] \times [x-(-13-8\sqrt{3})] = x^2+26x-23 = 0$.

Contests written and compiled by Steven R. Conrad, Daniel Flegler, & Adam Raichel   ©2018 by Mathematics Leagues Inc.

Problem 6-1

Since $2018 = 1 \times 2018 = 2 \times 1009$ is the product of two primes, the product of the 4 positive integer divisors of $2018 = 1 \times 2 \times 1009 \times 2018 = \boxed{2018^2}$.

Problem 6-2

We want to add together the three unit fractions with the smallest positive integer denominators. The sum is $\frac{1}{1} + \frac{1}{2} + \frac{1}{3} = \boxed{\frac{11}{6}}$.

Problem 6-3

Use the diagram at the right as a guide: the similarity of the larger and smaller rectangles tells us that $\frac{2a}{b} = \frac{b}{a}$, so $\frac{b}{a} = \boxed{\sqrt{2}}$.

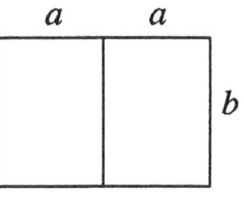

Problem 6-4

The smallest angle is opposite the shortest side.

Method I: The smallest angle is $\operatorname{Arc tan}(\frac{6}{8})$. The angle bisector's slope is the tangent of half that angle $= T = \tan(\frac{1}{2} \operatorname{Arc tan} \frac{3}{4})$. Since $\tan \frac{\theta}{2} = \frac{\sin \theta}{1 + \cos \theta}$, we get $T = \frac{3/5}{1 + 4/5} = \frac{1}{3}$. Inside the triangle, the 2 lattice points on the angle bisector are $\boxed{(3,2),\ (6,3)}$.

Method II: $\triangle I \cong \triangle II$ because their longest sides each have a length of 5, and the other two sides are congruent. The common side of the triangles is an angle bisector through $(3,2)$ and $(6,3)$.

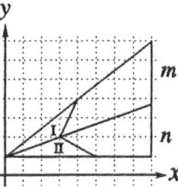

Method III: By the $\angle$ bisector theorem, $\frac{n}{m} = \frac{8}{10}$. From the graph, the slope $= \frac{1}{3}$. Continue as above.

Problem 6-5

It's a binomial distribution. The (# heads, freq.) pairs are $(0,1)$, $(1,4)$, $(2,6)$, $(3,4)$, $(4,1)$. The average of the squares is $\frac{1 \cdot 0^2 + 4 \cdot 1^2 + 6 \cdot 2^2 + 4 \cdot 3^2 + 1 \cdot 4^2}{1 + 4 + 6 + 4 + 1} = \frac{80}{16} = \boxed{5}$.

Problem 6-6

Since $z^3 = 1$, we have $\frac{1}{z} = z^2$, $\frac{1}{z^2} = z$, $\frac{1}{z^3} = 1$, and $P(z^2) = 2z^{20} + 3z^{18} + 4z^2 + 9$. With the substitution $z^3 = 1$, we transform the preceding equation into
* $P(z^2) = 2z^2 + 3 + 4z^2 + 9$. The very same substitution transforms $P(z) = 2z^{10} + 3z^9 + 4z + 9$ into
** $P(z) = 2z + 3 + 4z + 9$. Finally, $P(1)$ becomes
*** $P(1) = 2 + 3 + 4 + 9$, By adding together the three starred equations, we get $P(z^2) + P(z) + P(1) = 2(z^2 + z + 1) + (3 + 3 + 3) + 4(z^2 + z + 1) + (9 + 9 + 9)$. Since $z^3 - 1 = (z-1)(z^2 + z + 1) = 0$ and z is not real, so $z^2 + z + 1 = 0$. Therefore, we can write $P(z^2) + P(z) + P(1) = 2(0) + 9 + 4(0) + 27 = \boxed{36}$.

Contests written and compiled by Steven R. Conrad, Daniel Flegler, & Adam Raichel   ©2018 by Mathematics Leagues Inc.

Problem 1-1

The average of two numbers is half their sum, so the average here is 2018/2 = $\boxed{100}$.

Problem 1-2

The length of one leg of the right triangle in the diagram is 3, and the length of the hypotenuse is 5, so the length of the other leg is 4.

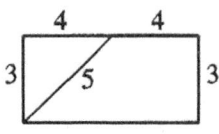

The length of the rectangle's upper side is $2 \times 4 = 8$. Since the length of the shorter side of the rectangle is 3, the area of the rectangle is $3 \times 8 = \boxed{2}$.

Problem 1-3

The minute hand has moved 180° past the 12. The hour hand has moved half the way from the 12 to the 1. That's $(1/2)(30°) = 15°$. The angle between the hour and minute hands is a $\boxed{165}$ angle.

Problem 1-4

Each smaller pile contains 2 cards. Since only 1 pile contains 1 card of each color, the other piles must each contain 2 red cards or 2 black cards. There must be an even number of cards of each color in these other piles. Since there is 1 pile with a single card of each color, the total number of cards of each color is an odd number. That odd number can be any of the $\boxed{1}$ odd numbers from 25 through 49.

Problem 1-5

If n is a positive integer that leaves a remainder of 24 when divided into 9449, then n must meet only two requirements: 1) $n > 24$, and 2) n must leave a remainder of 0 when divided into $9449 - 24 = 9425 = 5^2 \times 13 \times 29$. The 12 divisors of 9425 are 1, 5, 13, 5^2, 29, 5×13, 5×29, $5^2 \times 13$, 13×29, $5^2 \times 29$, $5 \times 13 \times 29$, and $5^2 \times 13 \times 29$. The first 3 divisors are smaller than 24, so the total number of divisors that meet both conditions is $\boxed{9}$.

[Note: If every letter represents a positive integer, and if $N = x^a y^b z^c$ is N's prime factorization, then N has $(a+1)(b+1)(c+1)$ different positive integer divisors.]

Problem 1-6

First label the positions 1, 2, 3, 4, 5, 6, and 7, and then assign the two girls to specific positions (with the boys filling in the remaining positions in any order). A "favorable" position is achieved only when girls choose positions 1 & 2 or 1 & 3 or 1 & 6 or 1 & 7 or 2 & 3 or 2 & 7 or 5 & 6 or 5 & 7 or 6 & 7. That's a total of 9 ways. The 2 girls can choose their positions from any of the 7 available positions in a total of 7 choose 2 = 21 ways. The required probability is $9/21 = \boxed{\frac{3}{7}}$.

Contests written and compiled by Steven R. Conrad, Daniel Flegler, & Adam Raichel   ©2018 by Mathematics Leagues Inc.

Problem 2-1

A square's sides have equal lengths, so $x-1 = 5-x$, and $x = 3$. The square's area is $(3-1)^2 = \boxed{4}$.

Problem 2-2

Since $4x^2+kxy+4y^2 = (2x\pm2y)^2 = 4x^2\pm8xy+4y^2$, either $k = 8$ or $k = -8$. Their sum is $\boxed{0}$.

Problem 2-3

If the smaller triangle has side-lengths x, x, and y, then the larger has side-lengths $2x$, $2x$, and y. Since $2x + y = 18$, while $4x + y = 28$, it follows that $x = 5$ and the length of the base $= y = \boxed{8}$.

Problem 2-4

The least common multiple of n and $1000 = 2^3 \times 5^3$ will be $2000 = 2^4 \times 5^3$ if and only if $n = 2^4 \times 5^m$, where m is an integer, $0 \le m \le 3$. The only possible values of n are $2^4 \times 5^0$, $2^4 \times 5^1$, $2^4 \times 5^2$, and $2^4 \times 5^3$. These four values of n are $\boxed{16, 80, 400, \text{ and } 2000}$.

Problem 2-5

Wherever a 20 appears as one of Jan's rank-numbers, a 1 appears in the same spot as one of Ann's rank-numbers (and *vice versa*). In fact, if two rank-numbers appear in the same position for Jan and for Ann, their sum is always 21; so the sum of Jan's first five rank-numbers and Ann's first five rank-numbers is $5 \times 21 = 105$. Since the sum of Jan's five numbers is 66, the sum of the Ann's five numbers is $105-66 = \boxed{39}$.

Problem 2-6

Let's try some. Which numbers can (and which can't) be written as a sum of consecutive positive integers? 1: no; 2: no; $3 = 1+2$; 4: no; $5 = 2+3$; $6 = 1+2+3$; $7 = 3+4$; 8: no; $9 = 4+5$; $10 = 1+2+3+4$; $11 = 5+6$; $12 = 3+4+5$; $13 = 6+7$; $14 = 2+3+4+5$; $15 = 7+8$; 16: no; $17 = 8+9$; . . . . The positive integers that aren't a sum of two or more consecutive positive integers are 1, 2, 4, 8, 16, ... These are all powers of 2, so the answer is probably the highest power of 2 less than 2018. That's 2^{10} or $\boxed{1024}$. Here's a proof.

Theorem: A positive integer n can be written as a sum of two or more consecutive positive integers if and only if n has an odd divisor > 1.

Proof: For $n > 0$, if $n = a+(a+1)+...+(a+k)$ for some positive integers a and k, the sum of the first and last terms is $2a+k$ and the number of terms is $k+1$, so $n = (2a+k)(k+1)/2$. If k is odd, then $2a+k$ is odd and $k+1$ is even; and if k is even, then $k+1$ is odd and $2a+k$ is even. Since both a and k are positive integers, both $2a+k$ and $k+1$ are greater than 1. Whichever of these is odd is an odd factor of n that is greater than 1. This proves that n is divisible by an odd number > 1. Let's reverse direction. If $n > 0$ has an odd divisor > 1 then $n = (2d+1)m$ for positive integers d and m. Therefore, we can write n as the sum $(m-d)+(m-d+1)+...+m+...+(m+d-1)+(m+d)=(2d+1)m = n$. Unless $m < d$, this is the sum of consecutive positive integers. If $m < d$, cancel every negative term with its corrsponding positive term, leaving a sum of consecutive terms which equals $(2d+1)m$. This indicated sum contains at least 2 positive integers, otherwise $m-d = -(m+d-1)$, implying that $m = 1/2$. The only positive integers which cannot be expressed as a sum of two or more consecutive positive integers are those which have no odd divisor greater than 1. The only such positive integers are the powers of 2.

Contests written and compiled by Steven R. Conrad, Daniel Flegler, & Adam Raichel   ©2018 by Mathematics Leagues Inc.

Problem 3-1

Since $n = (a-1)(a+1) = a^2-1 = 1000001^2-1$, we will get a perfect square if we add $\boxed{1}$.

Problem 3-2

Multiplying both sides by xy, we get $(x+y)xy = y+x$. Since $x+y \neq 0$, we can divide both side by $(x+y)$ to get $xy = \boxed{1}$.

Problem 3-3

If the first kid takes 1 bite, and all 2017 other kids take 2 bites, then the mean number of bites taken by all 2018 is slightly below the 2 bites taken by the other $\boxed{2017}$ kids.

Problem 3-4

If parallel lines each intersect a circle twice, the arcs between the lines are congruent. Since $\overset{\frown}{AP} \cong \overset{\frown}{BQ}$, $m\angle BOQ = m\angle AOP = 30$.

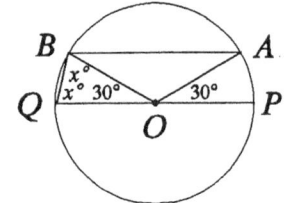

Since $\triangle BOQ$ is an isosceles triangle with a vertex angle of 30°, each of its base angles is a $\boxed{75°}$ angle.

Problem 3-5

Let the first term be a and let the common ratio be r. Since the sum of the first six terms is 4, and the sum of the first three terms is 3, we have that $a+ar+ar^2+ar^3+ar^4+ar^5 = 4 = 3+r^3(a+ar+ar^2) = 3 + 3r^3$, so $r^3 = 1/3$. The sum of the first twelve terms is $4 + r^6(a+ar+ar^2+ar^3+ar^4+ar^5) = 4+(\frac{1}{9})(4) = \boxed{40/9}$.

Problem 3-6

Each of the first 3 diagrams at the right contains a shaded region whose area is 1/6 of the hexagon's area. In the first diagram, one of 6 congruent equilateral triangles with a vertex at the center is shaded. In the second diagram, $\triangle ABC$ and $\triangle BOC$ are both half of $\square BCOA$, so $\triangle ABC$ is 1/6 of the hexagon. In the third diagram, compare the areas of triangles ACM and COD. Although $CD = 2 \times AM$, the altitude from O to $\overline{CD}$ is half the length of the altitude from C to $\overline{AM}$, so the triangles are equal in area. In the final diagram, quadrilateral ABCM has an area of 24. Since that's the sum of the areas of $\triangle ACM$ and $\triangle ABC$, 24 is 1/3 of the hexagon's area. The area of the shaded region is $3 \times 24 - 31 = \boxed{41}$.

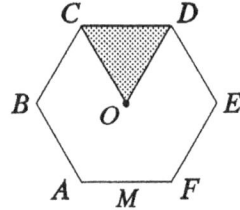
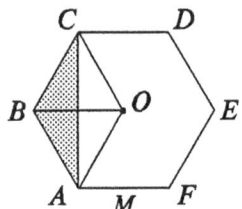
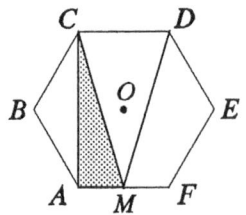
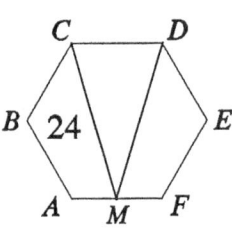

Contests written and compiled by Steven R. Conrad, Daniel Flegler, & Adam Raichel   ©2018 by Mathematics Leagues Inc.

Problem 4-1

Since $\sqrt{x^2} = |x|$, we must solve $|x| = 2019$. The two solutions are $x = \boxed{\pm 2019}$.

Problem 4-2

In an isosceles right triangle, if each leg's length is x, then the length of the hypotenuse is $x\sqrt{2}$. The area of the triangle is $x^2/2 = 4$, so $x\sqrt{2} = \boxed{4}$.

Problem 4-3

Connect the circle's center to the endpoints of the chord of length 10, as seen. Since $\pi r^2 = 100\pi$, the circle's radius-length is 10. Hence, the triangle just drawn is an equilateral triangle. Its central angle and its intercepted arc are both 60°. An inscribed angle that intercepts a 60° arc has a degree-measure of $\boxed{30}$.

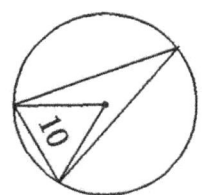

Problem 4-4

The average of 25 consecutive integers is always the middle term of the sequence, so write the sequence as $n-12, n-11, \ldots, n, \ldots, n+11, n+12$. The middle term, n, equals the average term, so $(n-12)^2 = n$, from which $n = 9$ or 16. The least possible value of the last term of the sequence is $n+12 = \boxed{21}$.

Problem 4-5

Since $3^a = 9^b$ implies $3^a = 3^{2b}$, $a = 2b$, and $\frac{a}{b} = 2$.

Since $9^b = 27^c$ implies $3^{2b} = 3^{3c}$, $2b = 3c$, so $\frac{b}{c} = \frac{3}{2}$.

Since $3^a = 27^c$ implies $a = 3c$, $\frac{c}{a} = \frac{1}{3}$.

Finally $\frac{a}{b} + \frac{b}{c} + \frac{c}{a} = 2 + \frac{3}{2} + \frac{1}{3} = \boxed{\frac{23}{6}}$.

Problem 4-6

If the reciprocals are written with a common denominator, the numerators are an arithmetic sequence. Any such sequence can be formed by taking any 5 different positive integers in arithmetic progression, finding their lcm, and forming the 5 fractions with each integer as a numerator and the lcm as denominator. For example, try 1, 2, 3, 4, 5 with lcm 60. The unit fractions are 5/60 = 1/12, 4/60 = 1/15, 3/60 = 1/20, 2/60 = 1/30, and 1/60. Next try 2, 3, 4, 5, 6. The lcm is 60 again, and the unit fractions are 1/10, 1/12, 1/15, 1/20, 1/30. If we multiply any set of 5 integers by a positive constant, the new set forms the identical sequence of unit fractions. If we consider larger sets of 5 different integers (with gcd = 1) in arithmetic progression, the lcm > 60 and the denominators of the fractions have larger sums. (Below we show when n is the largest integer in a set of 5 such integers, the lcm is $\geq 10n$ and that if $n \geq 7$, the lcm > $30n$. Thus, the fractions whose denominators have the least possible sum are 1/10, 1/12, 1/15, 1/20, 1/30. Hence the least such sum is $10+12+15+20+30 = \boxed{87}$.

PROOF that lcm $\geq 10n$: assume without loss of generality that the gcd of the 5 integers is 1. Let n, $n-d$, $n-2d$, $n-3d$, $n-4d$ be the 5 integers, with d relatively prime to each of them. The 2 largest are relatively prime, so lcm must be $\geq n \times (n-d)$. The third largest integer is relatively prime to the second largest, but might share a factor of 2 with the largest since they differ by $2d$. Thus, lcm $\geq n \times (n-d) \times (n-2d)/2$. Since $n-4d \geq 1$, $d \leq (n-1)/4$. Substituting for d in the previous inequality, we have lcm $\geq n \times (n+1) \times (3n+1)/16$. The lcm is an increasing function of n. If $n \geq 7$, lcm $\geq 11n$. If $n = 5$ or 6, we saw above that lcm $\geq 10n$.

PROOF that lcm > $30n$ for $n \geq 7$: for $n \geq 12$, the above approximation shows that lcm > $30n$. Therefore, the largest unit fraction is less than 1/30. By direct calculation, lcm for $n = 7,8,9,10,11$ is also > $30n$.

Problem 5-1

Since making N's hundreds' digit a 9 and its units' digit a 0 will maximize the difference, no matter what the middle digit is, $N - M = 9\underline{?}0 - 0\underline{?}9 = \boxed{891}$.

Problem 5-2

Since $ab = 5$, and a and b are integers, (a,b) must be one of $(1,5)$, $(5,1)$, $(-1,-5)$, or $(-5,-1)$. The least value of a^b is $(-1)^{-5} = \boxed{-1}$.

Problem 5-3

By trial and error, take each positive perfect square and subtract 1 to find every positive integer n for which $n+1$ is a perfect square. Find the least of those integers n for which $2n+1$ is the square of an integer. Let's try 1. That fails because $1-1 = 0$ is not positive. Try 4: $4-1 = 3$, but $2(3)+1 = 7$, which is not a perfect square. We try 9, then 16, and when we finally try 25, we get $25-1 = 24$, and $2(24)+1 = 49$. Both $n+1$ and $2n+1$ are perfect squares when $n = \boxed{24}$.

Problem 5-4

If B is the number of Borgs, and C is the number of Corgs, then the number of Corgs is $2B+13$. Therefore, $B+(2B+13) = 100$, $B = 29$, and $C = \boxed{71}$.

Problem 5-5

Whenever $a > 1$ and $b > 1$, $\log_a b = \dfrac{1}{\log_b a}$. We can write $\log_x 4 + \log_x 9 + \log_x 16 + \ldots + \log_x 2019^2 = k$, or $\log_x(2^2 \times 3^2 \times 4^2 \times \ldots \times 2019^2) = k$. It follows that $x^k = (2^2 \times 3^2 \times 4^2 \times \ldots \times 2019^2) = (2019!)^2$, so $k = \boxed{2}$.

Problem 5-6

Method I: The square's side is $8\sqrt{3}$, so its diagonal is $8\sqrt{6} = d$. The trapezoid's height is $\dfrac{d}{4}$. Its longer base is $\dfrac{d}{2}$. The shaded $\triangle$'s legs trisect the diagonal (to prove this, use similar triangles I and II outlined in the middle diagram), so the shorter base is $\dfrac{d}{3}$. The trapezoid's area is $\dfrac{h}{2}(b_1+b_2) = \dfrac{d}{8}(\dfrac{d}{2}+\dfrac{d}{3}) = \dfrac{5d^2}{48} = \boxed{40}$.

Method II: Two of the shaded triangle's vertices are midpoints of sides of the square, so the smaller unshaded region is 1/8 of the square, and the other unshaded regions are each 1/4 of the square. The shaded triangle's area is $(3/8) \times 192 = 72$. 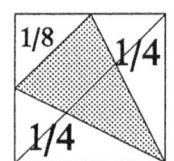 Notice that $\triangle$ I is similar to $\triangle$ II. Since M is a midpoint, the parallel sides of $\triangle$s I and II are in the ratio $x:2x = 1:2$, as are all corresponding 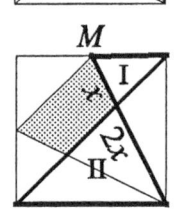 sides. Thus, the smaller shaded triangle and the larger shaded triangle are similar, with ratio of similitude $2k:3k = 2:3$. Since the area ratio of similar triangles is the square of their ratio of similitude, the ratio of the areas of the shaded triangles is $(2/3)^2 = 4:9$. The smaller shaded 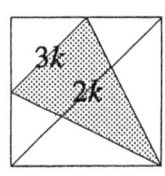 triangle's area $= (4/9) \times 72 = 32$. The trapezoid's area is what remains: $72-32 = 40$.

Contests written and compiled by Steven R. Conrad, Daniel Flegler, & Adam Raichel   ©2019 by Mathematics Leagues Inc.

Contest # 6   *Answers & Solutions*   **3/19/19**

Problem 6-1

As seen in the diagram at the right, if we let length of the shorter side of each of the congruent rectangles be x, then the length of each side of the square is $3x$. The perimeter of each of the 3 rectangles is $8x = 40$, so $x = 5$ and the perimeter of the square $= 12x = \boxed{60}$.

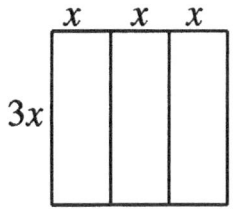

Problem 6-2

Let's check each integer greater than 237 until we get the right answer. Neither 238 nor 239 is correct, because neither number's units digit is a prime. No number from 240 through 249 is correct because 4 is not a prime. Neither 0 nor 1 is a prime, so neither 250 nor 251 is correct. The smallest integer greater than 237 each of whose digits is a prime is $\boxed{252}$.

Problem 6-3

The sum of the 16 integers is 120. If the sum of the numbers in set A is x, then the sum of the numbers in sets B, C, and D are respectively $2x$, $4x$, and $5x$. The sum of all four sums is $12x = 120$, so $x = \boxed{10}$.

[Such sets do exist. One example is $A = \{0,1,3,6\}$, $B = \{2,4,5,9\}$, $C = \{7,10,11,12\}$, $D = \{8,13,14,15\}$.]

Problem 6-4

Since one endpoint is at $(a,0)$ and the midpoint is $(2019,2020)$, the endpoint that lies on the line $y = x$ is $(4038-a,4040)$. Since these two coordinates are equal, it follows that $a = \boxed{-2}$.

Problem 6-5

Adding and subtracting the two given equations, we get $\cos x \cos y + \sin x \sin y = 1/2$, or $\cos(x-y) = 1/2$ and $\cos x \cos y - \sin x \sin y = 1$, or $\cos(x+y) = 1$. We want the solution(s) with the least absolute values for x and y. The minimal absolute value solutions arise from solving $x-y = \pm\pi/3$ and $x+y = 0$. The two solutions are $\boxed{(\pi/6, -\pi/6), \ (-\pi/6, \pi/6)}$.

Problem 6-6

Every such triangle's vertices lie inside the circle, so any two sides of such a triangle lie on two chords that intersect inside the circle. The non-concurrency of any three chords inside the circle has an important consequence: no matter which 6 of the 10 points are chosen, there's exactly one way these 6 points can and do serve as endpoints of three chords whose intersections determine a triangle wholly inside the circle, as seen below. The number of such triangles is $_{10}C_6 = \boxed{210}$.

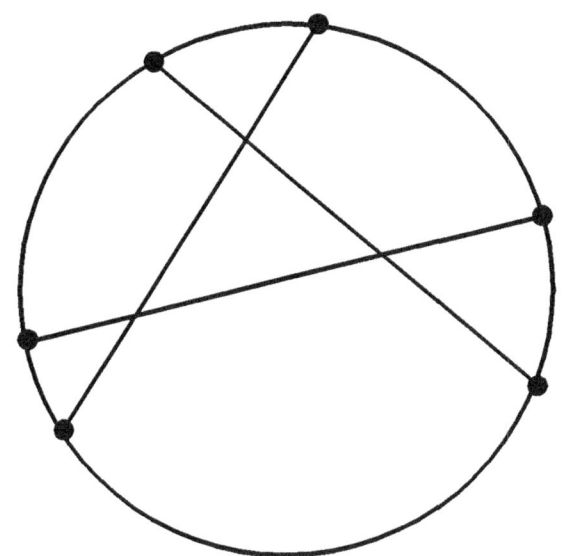

Contests written and compiled by Steven R. Conrad, Daniel Flegler, & Adam Raichel   ©2019 by Mathematics Leagues Inc.

Problem 1-1

The sum of two odd primes is even. Here, the sum of the two primes is 2019, an odd number, so one of the primes is even. The only even prime is 2, so the other prime is $2019 - 2 = \boxed{201}$.

Problem 1-2

The length of each diagonal of the 3×4 rectangle is 5. The side-lengths of each small right triangle are half the side-lengths of the 3-4-5 triangle. Since the 3-4-5 triangle has perimeter 12, the perimeter of each small right triangle is half that, $\boxed{6}$.

Problem 1-3

If all 3 integers were 0, their sum would be 0. If all 3 were 9, their sum would be 27. Therefore, the number of different possible sums of all 3 integers is $\boxed{2}$.

Problem 1-4

The length of the base must be 22, and the length of each leg must be 61. In a right triangle with a hypotenuse of length 61 and a leg of length 11, the length of the other leg (the isosceles triangle's altitude) is 60. The area of the triangle is $11 \times 60 = \boxed{66}$.

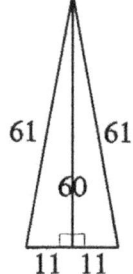

Problem 1-5

The repetition of "97" implies that $9797 = 101 \times 97$. Now, let's write each factor as a sum of 2 squares. Since $101 = 10^2 + 1^2$, we can use $(a,b) = (10,1)$ or $(1,10)$. Similarly, since $97 = 9^2 + 4^2$, we could use $(c,d) = (9,4)$ or $(4,9)$. Hence, $(ac+bd)^2 + (ad-bc)^2 = (90+4)^2 + (40-9)^2 = 94^2 + 31^2$. If we use $(c,d) = (4,9)$, then we get $49^2 + 86^2$ instead. The four possible answers are $\boxed{(31,94),\ (94,31),\ (49,86),\ \text{or}\ (86,49)}$.

Any ONE answer gets FULL CREDIT.

Parentheses are NOT required.

[Note: $(ac-bd)^2 + (ad+bc)^2$ can also be used.]

Problem 1-6

With 2 choices (yes or no) for each topping, it follows that, with 6 toppings, there are a total of $2^6 = 64$ different combinations of toppings (from no topping to all 6 toppings) on each pizza. To answer the question, there are only 3 cases to consider: **1)** the 3 pizzas can have the same selection of toppings in 64 different ways; or **2)** the 3 pizzas can have completely different selections of toppings in $\binom{64}{3} = (64 \times 63 \times 62)/3!$ ways; or **3)** 2 pizzas can have the same toppings, different from the third pizza. To count the number of pizzas in this case, start with 2 different topping selections. The third pizza must match either of the other 2, so the number of ways this third case can happen is $2 \times \binom{64}{2}$. Adding, we find that the total number of ways from all 3 cases is $\boxed{45\,760}$.

Contests written and compiled by Steven R. Conrad, Daniel Flegler, & Adam Raichel   ©2019 by Mathematics Leagues Inc.

Problem 2-1

If $\dfrac{x-1}{x+1} = 2019$, then $\dfrac{x^2-2x+1}{x^2+2x+1} = \left(\dfrac{x-1}{x+1}\right)^2 =$ $\boxed{(2019)^2 \text{ or } 4\,076\,361}$.

Problem 2-2

By trial and error, $4+9 = \boxed{13}$.

Problem 2-3

As seen in the diagram drawn below, the number of such points is $\boxed{20}$.

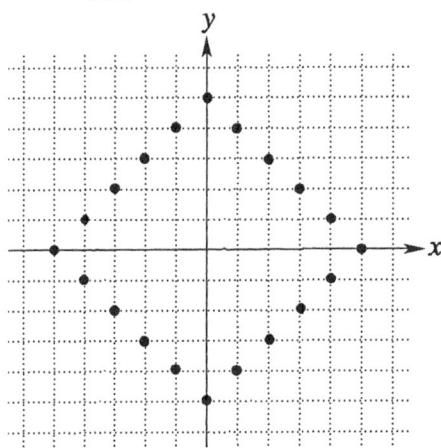

Problem 2-4

By connecting the interior point to the vertices of the triangle (shown as dotted lines in the diagram), 3 obtuse triangles are created. If the equilateral triangle's side is s and its altitude is h, and if the

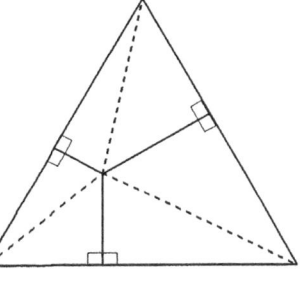

distances from the interior point to the sides are d_1, d_2, and d_3, we get $(1/2)sh = (1/2)sd_1 + (1/2)sd_2 + (1/2)sd_3$. Therefore, the length of the altitude of the equilateral triangle is $d_1+d_2+d_3$, the sum of the three distances. An equilateral triangle with altitude $11\sqrt{3}$ has an area of $\boxed{121\sqrt{3}}$.

Problem 2-5

Expand the right side of the given equation to get $(x-c)(x-c)(x+\frac{a}{c^2}) = x^3+x^2(\frac{a}{c^2}-2c)+x(c^2-\frac{2a}{c})+a = x^3+cx^2-5x+a$. Next, let's equate coefficients of x^2. We get $\frac{a}{c^2}-2c = c$, from which $a = 3c^3$. Equating coefficients of x, $c^2-\frac{2a}{c} = -5$, so $a = \frac{c^3+5c}{2}$. Since $a = 3c^3 = \frac{c^3+5c}{2}$, $c^2 = 1$, $c = \pm 1$, and $a = \boxed{\pm 3}$.

Problem 2-6

Every real number x is composed of an integer part, $[x]$, and a fractional part, f, $0 \le f < 1$. Note that $f = 0$ when the real number is an integer. Thus, $x-[x] = f$. From the given information, $\sqrt{x} = x-[x]+2 = f+2$. Thus, $[x]+f = x = (\sqrt{x})^2 = (f+2)^2 = f^2+4f+4$, so $[x] = f^2+3f+4$. As f increases towards 1, the right side of the previous equation increases towards 8, but never reaches 8. Therefore, the largest possible value of $[x]$ is 7. If we replace $[x]$ with 7 in the previous equation, we can solve for f, and then use $x = [x]+f$ to solve for x. To maximize f, solve $7 = f^2+3f+4$, whose one solution between 0 and 1 is $f = \frac{-3+\sqrt{21}}{2}$. Thus, $x = 7+f = 7+\frac{-3+\sqrt{21}}{2} = \boxed{\frac{11+\sqrt{21}}{2}}$.

[**NOTE**: For this question, only an exact answer can get credit. An approximation will get no credit.]

Contests written and compiled by Steven R. Conrad, Daniel Flegler, & Adam Raichel   ©2019 by Mathematics Leagues Inc.

Problem 3-1

Method I: The number midway between 2019 and 2020, $\boxed{2019.5}$, satisfies the given equation.

Method II: $x^2-4038x+2019^2 = x^2-4040x+2020^2$, so $2x = 2020^2 - 2019^2 = 4039$, and $x = 2019.5$.

Problem 3-2

If there were no restriction against selecting consecutive integers for consecutive side-lengths, then the maximum would be $4+5+6 = 15$. With the restriction, if we choose side-lengths 6,3,5,1,4,2, we get the maximum restricted side-length sum of $\boxed{14}$.

[**NOTE:** The largest side being smaller than the sum of the other sides is a necessary and sufficient condition for a polygon to exist. This is true whether the order of the sides is given or not.]

Problem 3-3

Method I: Add consecutive integers. Set the sum equal to 1000. Let's begin: $x + x+1 = 1000$ (no integer solution); $x + x+1 + x+2 = 1000$ (no integer solution); $x + x+1 + x+2 + x+3 = 1000$ (no integer solution); $x + x+1 + x+2 + x+3 + x+4 = 1000$ has the solution $x = 198$. The number of consecutive integers needed is at least $\boxed{5}$.

Method II: If k were even, then the sum of the consecutive integers wold be k time the average of the two middle integers. To get a total of 1000, if k were 2, then the average of two consecutive integers would have to be 500; if k were 4, then the average of two consecutive integers would have to be 250. Neither is possible. If k were odd, then the sum of the consecutive integers would be k times the middle integer/ Thus, k canot be 3, since 1000 isn't divisible by 3, but k can be 5, since $5 \times 200 = 1000$. The consecutive integers would be $198+199+200+201+202$.

Problem 3-4

If we let $x = 2t+7$, then $t = (x-7)/2$ and $f(x) = 12t+37 = (12x-84)/2 + 37 = 6x-5$. Therefore, to solve $f(x) = x^2$, we must solve $6x-5 = x^2$. Factor to get $(x-1)(x-5) = 0$, from which $x = \boxed{1, 5}$.

Problem 3-5

If r is the length of a radius of each circle, then the length of the diagonal of the small square in the upper left-hand corner of the large square is

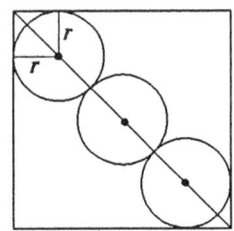

$r\sqrt{2}$. The length of the diagonal of the large square is therefore $4r+2r\sqrt{2}$. Since that diagonal's length is $8\sqrt{2}$, it follows that $4r+2r\sqrt{2} = 8\sqrt{2}$. Solving, $r = \boxed{\dfrac{4\sqrt{2}}{2+\sqrt{2}} \text{ or } 4\sqrt{2}-4}$.

Problem 3-6

To have each grandchild receive at least \$10, start by giving each \$9 and then count the number of ways to distribute the remaining \$55 so that each grandchild gets at least 1 of the \$1 bills. Think of those 55 \$1 bills as lined up on a table with a space between each dollar bill and the next one. There are 54 spaces between the 55 \$1 bills. Choose any 4 spaces to split 55 into 5 integers. (For example, selecting the 1st, 2nd, 23rd, and 50th spaces splits 55 into 1, 1, 21, 27, and 5.) The number of ways we can make such a selection is $_{54}C_4 = \boxed{316\,251}$.

Contests written and compiled by Steven R. Conrad, Daniel Flegler, & Adam Raichel   ©2019 by Mathematics Leagues Inc.

Problem 4-1

$\dfrac{(x-2019)(x+2019)}{x-2019} = x+2019 = 2020$, so $x = \boxed{1}$.

Problem 4-2

Start working down from 100, writing the primes until two are seaparated by 4. The primes are 97, 89, 83, 79, . . . , and that is all that is needed. Since 83 and 79 differ by 4, the answer is $83 + 79 = \boxed{162}$.

Problem 4-3

Since $2^{22}-2$ is divisible by 2 (the smallest prime), the largest proper divisor of $2^{22}-2$ is $(2^{22}-2)/2 = 2^{21}-1$, so $k = \boxed{21}$.

Problem 4-4

Since the base and altitude of both large unshaded triangles are the same as the base and altitude of the parallelogram, the area of each unshaded large triangle is 22, half the area of the parallelogram. Since the shaded region's total area is 14, the unshaded region's total area is 30. Remove either large unshaded triangle from the total unshaded region to show that the area of each small unshaded triangle is $30-22 = 8$. The unshaded region common to the large unshaded triangles has an area of $30-8-8 = \boxed{14}$.

[**NOTE:** It can be proved that the unshaded common region always has the same area as the shaded region.]

Problem 4-5

This question could be asked about any number of positive integers $n \geq 3$ having the same sum and product. A solution that would work in any case is to make all the numbers 1 except the last two, which would be 2 and n, resulting in a sum and product that are each $2n$. When $n = 24$, the sum and product is $\boxed{48}$.

[**NOTE:** Other solutions sometimes exist, but no other solutions exist for $n = 2, 3, 4, 6, 24, 114,$ or 174.]

Problem 4-6

The Angle Bisector Theorem tells us that the bisector of an angle of a triangle splits the opposite side in proportion to the other two sides. The vertical altitude of the isosceles

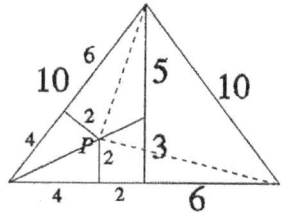

triangle has length 8 and is split by the angle bisector in the same ratio as the other two sides of the left-side triangle $= (4+6):(4+2) = 10:6 = 5:3$, as shown. Point P, whose distance from two sides is 2, is shown. The longer legs of the small right triangles are 4, as shown. Here's why: at the lower left, the right triangle with shorter leg 3 and longer leg 6 is similar to the right triangle with shorter leg 2, so the latter triangle's longer leg is 4, and the overall base of length 12 is split into segments of length 4, 2, and 6. By the Pythagorean Theorem, the sum of the squares of the distances from point P to the 3 vertices of the large triangle is $(2^2+6^2) + (2^2+4^2) + (2^2+8^2) = \boxed{128}$.

Contests written and compiled by Steven R. Conrad, Daniel Flegler, & Adam Raichel ©2020 by Mathematics Leagues Inc.

Problem 5-1

The solutions are -1, -2, and -3. The sum of their reciprocals is $-1 - \dfrac{1}{2} - \dfrac{1}{3} = \boxed{\dfrac{-11}{6}}$.

Problem 5-2

In any triangle, the length of the longest side must be less than the sum of the lengths of the other two sides, so $2020+n < 2020-n + 2020$, and $n < 1010$. The largest possible integer value of n is $\boxed{1009}$ (and such a triangle does exist).

Problem 5-3

This is basically a pigeon hole question. If there are 100 students and the only mode is 0, then we must make sure that no score occurs more often than a score of 0. Divide 100 by 16 (the number of possible scores) to get 6 (remainder 4). Thus, if each score is earned at most 7 times, we'd have 4 scores of 7 and 12 scores of 6. This doesn't allow for only one mode. If one score occurs 8 times, 2 scores occur 7 times, and the remaining 13 scores occur 6 times, we'd have accounted for all 100 students, and the only mode would be the score that occurred $\boxed{8}$ times.

Problem 5-4

The vertex angles of the two isosceles triangles are supplementary, so the 4 base angles of the isosceles triangles

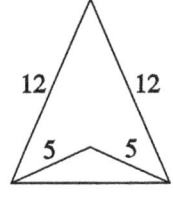 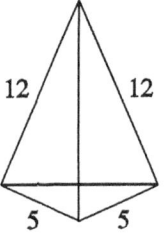

have a sum of 180°. Reflecting the smaller isosceles triangle across the common base creates two congruent right triangles with legs of length 5 and 12. The two altitudes together make up the common hypotenuse of the two right triangles. Its length is $\boxed{13}$.

Problem 5-5

Clearly, the numbers in question must have 3 digits each. Call one of them ABC. Call the other CBA. Since $A \times C$ ends in a 5, one of the numbers A or C—let's say its A—must be 5. Since $92{,}565 \div 500 < 200$, $C = 1$. To determine the value of B we notice that the 6 in the product is the last digit of $5B+B = 6B$, so $B = 1$ or $B = 6$. All that remains is to test the 2 possibilities, $\boxed{165 \ \& \ 561}$ (which work).

Problem 5-6

In a geometric sequence with common ratio r and first term a, the first 3 terms are a, ar, and ar^2. We are told that $a+4$, ar, and ar^2 form an arithmetic sequence, so $r \neq 1$ and $ar^2 - ar = ar - (a+4) = ar - a - 4$. This leads to $a(r^2 - 2r + 1) = -4$, so $a = \dfrac{-4}{(r-1)^2}$. Since a and r are integers, $(r-1)^2$ is a positive divisor of -4. The equation $(r-1)^2 = 2$ has no rational solution, so the non-zero solutions of $(r-1)^2 = 1$ or 4 are the only possibilities. From these two equations, we find that $r = 2$, 3, or -1. The three corresponding (a,b,c) triples are: $\boxed{(-4,-8,-16), \ (-1,-3,-9), \ (-1,1,-1)}$.

Contests written and compiled by Steven R. Conrad, Daniel Flegler, & Adam Raichel   ©2020 by Mathematics Leagues Inc.

Problem 6-1

Since 99999/11 leaves a remainder of 9, the largest 5-digit integer divisible by 11 is $99999 - 9 = \boxed{99990}$.

Problem 6-2

What is the largest perfect square less than 2020? Since $\sqrt{2020} \approx 44.9$, the largest perfect square less than 2020 is $44^2 = 1936$. Therefore, $2020 - n = 1936$, and $n = \boxed{84}$.

Problem 6-3

Of the first 8 tosses, exactly 1 will be "heads"; and the ninth toss must also be "heads." Only 8 permutations of the first 8 tosses contain 1 "heads" (7 "tails" and 1 "heads" can differ only in which toss is "heads"). The probability is $8 \times (\frac{1}{2})^1 (\frac{1}{2})^7 (\frac{1}{2}) = \frac{1}{2^6} = \boxed{\frac{1}{64}}$.

Problem 6-4

Method I:

From the area of 2 quarter-circles (each with radius 2 and area $a + 2b$), subtract the area of the square, $2a + 2b$. The result is $2b$, the shaded region's area. Its area is $\pi + \pi - 4 = \boxed{2\pi - 4}$.

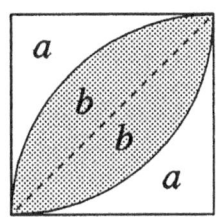

Method II:

A quarter-circle $(a + 2b)$ minus a right triangle $(a + b)$ is $b = \pi - 2$. Double that to get $2b = 2\pi - 4$, the area of the shaded region.

Problem 6-5

If $Y = 200y$ & $X = 200x$, then $Y = 200y = 200x^2 = 200\left(\frac{X}{200}\right)^2 = \frac{1}{200}X^2$, so $a = \boxed{\frac{1}{200}}$.

Problem 6-6

If $x = \log_4 a = \log_{10} b = \log_{25}(a+b)$, we can conclude that $4^x = a$, $10^x = b$ and $25^x = a+b$. It follows that $4^x + 10^x = 25^x = a+b$, and $\frac{a}{b} = \frac{4^x}{10^x} = \left(\frac{2}{5}\right)^x$. Also, $\frac{b}{a+b} = \frac{10^x}{25^x} = \left(\frac{2}{5}\right)^x$. Therefore, $\frac{a}{b} = \frac{b}{a+b}$, or $\frac{b}{a} = \frac{a}{b} + 1$. Now, if we let $c = \frac{a}{b}$, then we get that $\frac{1}{c} = c + 1$. The positive value of $c = \frac{a}{b}$ is the positive solution of $c^2 + c - 1 = 0$. That value of $c = \frac{a}{b}$ is $\boxed{\dfrac{-1 + \sqrt{5}}{2}}$.

[**NOTE:** The value of x is $\log_{2/5}(c)$, which, to seven decimal places, is $0.5251737\ldots$.]

Contests written and compiled by Steven R. Conrad, Daniel Flegler, & Adam Raichel   ©2020 by Mathematics Leagues Inc.

Problem 1-1

If we call the number x, then the problem statement says that $1 = 2020x$. Solving, $x = \boxed{\dfrac{1}{2020}}$.

Problem 1-2

Since the primes n and $n+2$ differ by 2, n is odd, so $n+2$ and $n+4$ are both odd. If an odd integer is divided by 3, the only possible remainders are 0, 1, or 2. If n leaves a remainder of 0, n would be the one divisible by 3. If n leaves a remainder of 1 when divided by 3, then $n+2$ would be divisible by 3. If n leaves a remainder of 2 when divided by 3, then $n+4$ would be divisible by 3. The only prime divisible by 3 is 3. Since 1 isn't prime, the only such sequence of primes is 3, 5, 7—so there's only $\boxed{1}$ value of n.

Problem 1-3

Method I: Coordinatize, placing the rectangle's lower left vertex at $(0,0)$. The length of the dashed line segment with endpoints $(6,0)$ and $(21,8)$ is $\boxed{1}$.

Method II:

Solution without words:

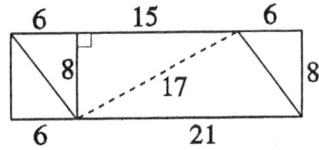

Problem 1-4

In the employee-intern pairings, equal numbers of employees and interns are selected. Of those who remain, there are as many employees as interns, so the probability that the number of copper rings equals the number of brass rings is $\boxed{1}$.

Problem 1-5

Call the number N. The sum of N's two largest divisors is odd, so one of N's two largest divisors is even, hence 2 is also a divisor of N. Since 2 is clearly N's smallest divisor > 1, N's largest proper divisor is $N/2$. Since N's largest divisor is N itself, $N + N/2 = 111$, so $N = \boxed{74}$.

Problem 1-6

Since a and b have greatest common divisor 1, the highest power of each prime divisor of ab must be a factor of a or a factor of b, but not of both. The prime factors of 30! are the primes below 30, and the highest power of each such prime will be a factor of exactly one of a or b. The primes less than 30 are 2, 3, 5, 7, 11, 13, 17, 19, 23, and 29, ten primes in all. Imagine two jars, one labeled a and one labeled b. After we put a 1 into each jar, each of the 10 primes (including it powers) has two possible landing places: the jar labeled a or the jar labeled b. Hence, the total number of possible integer pairs (a,b) is 2^{10}. In half of these pairs, $a < b$; in half, $a > b$. The answer to the question is $2^{10}/2 = 2^9$ or $\boxed{512}$.

[**NOTE:** By factoring into primes, we get $30! = 2^{26} \times 3^{14} \times 5^7 \times 7^4 \times 11^2 \times 13^2 \times 17^1 \times 19^1 \times 23^1 \times 29^1$.]

Contests written and compiled by Steven R. Conrad, Daniel Flegler, & Adam Raichel   © 2020 by Mathematics Leagues Inc.

Problem 2-1

Since 1 is not a prime, the three smallest primes are 2, 3, and 5. Their sum is 10. The first square after 10 is 16, and $2 + 3 + 11 = \boxed{16}$.

Problem 2-2

Gerry arrived x hours past noon, Dale arrived 4 hours later, and x hours later, it was 5 PM, so $2x + 4 = 5$. Thus, $x = 0.5$, so Gerry arrived at $\boxed{\text{12:30 PM}}$.

Problem 2-3

Since $\dfrac{x^2 + 2021x + 2020}{x^2 - 2020x - 2021} = \dfrac{(x + 2020)(x + 1)}{(x - 2021)(x + 1)} = 2$, and since $x \neq -1$, it follows that $x + 2020 = 2(x - 2021)$, from which $x = \boxed{6062}$.

Problem 2-4

The first four odd squares are 1, 9, 25, and 49. Their differences are 8, 16, 24, 40, and 48. All are multiples of 8. Now pick two odd numbers $2x+1$ and $2y+1$. If $2x+1 > 2y+1$, then the difference of their squares factors into $4(x-y)(x+y+1)$. Since $x-y$ and $x+y+1$ have opposite parities, one of them must be even (hence divisible by 2). Therefore, the difference of their square factors is divisible by $4 \times 2 = 8$. Since $3^2 - 1^2 = 8$, the greatest integer divisor is $4 \times 2 = \boxed{8}$.

Problem 2-5

Every multiple of 1000 is divisible by 8, so a 4-digit number is divisible by 8 if, after removing its thousands digit, the remaining 3-digit number is divisible by 8. From 500 to 599, the 12 numbers 504, 512, 520, 528, 536, 544, 552, 560, 568, 576, 584, and 592 are the only ones divisible by 8. For each thousands digit, there are 12 numbers with hundreds digit 5 that are divisible by 8. Therefore, the total number of such numbers from 1000 to 9999 is $9 \times 12 = \boxed{108}$.

Problem 2-6

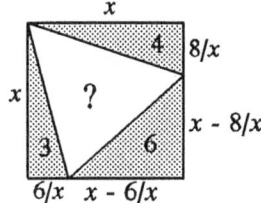

Call the length of one side of the square x. Since the areas of the two smaller shaded triangles are 3 and 4, the shorter legs of these right triangles have respective lengths $6/x$ and $8/x$. Subtracting from x, the respective lengths of the legs of the large shaded right triangle are $x - (6/x) = (x^2 - 6)/x$ and $x - (8/x) = (x^2 - 8)/x$. The product of these legs is twice the area of the largest shaded triangle, so it follows that $(x^2 - 6)(x^2 - 8)/x^2 = 12$. Clear fractions, expand, and factor, and you'll get $x^4 - 26x^2 + 48 = 0 = (x^2 - 24)(x^2 - 2)$. Either $x^2 = 2$ or $x^2 = 24$. Since the area of the square is x^2, and the area of the square exceeds 2, it follows that $x^2 = 24$. Finally, the area of the unshaded triangle is $24 - (3 + 6 + 4) = \boxed{11}$.

Contests written and compiled by Steven R. Conrad, Daniel Flegler, & Adam Raichel   ©2020 by Mathematics Leagues Inc.

Problem 3-1

No Pythagorean triple has hypotenuse 18 or 19. The 12–16–20 (or 16–12–20) triple has the property we seek. That day can be written as $\boxed{\text{Dec. 16, 2020}}$ or as either 12–16–20 or 16–12–20.

Problem 3-2

A fraction like $a + \cfrac{1}{b + \cfrac{1}{c + \cfrac{1}{d}}}$ is called a *continued frac-*

tion. Use a "split and flip" approach. Split off the integer part to get $46/35 = 1 + 11/35$. Now flip the fraction to get $46/35 = 1 + 1/(35/11)$. Repeat the procedure with $35/11 = 3 + 2/11 = 3 + 1/(11/2)$. Finally, since $11/2 = 5 + 1/2$, the complete expansion is $46/35 = 1 + \cfrac{1}{3 + \cfrac{1}{5 + \cfrac{1}{2}}}$, from which $d = \boxed{2}$.

Problem 3-3

The first 5 palindromes greater than 9 999 are 10 001, 10 101, 10 201, 10 301, and $\boxed{10\,401}$.

Problem 3-4

Let $y = \sqrt[3]{x}$, from which $x = y^3$. Substituting into the given equation, $(y^3)^y = (y^3)(y)$, or $y^{3y} = y^4$. Either both bases are equal to 1 or the exponents are equal. If $y = 1$, then $x = 1$. If not, then $3y = 4$, and $y = 4/3$, from which $x = 64/27$. Both check, so the two possible values of x are $\boxed{\dfrac{64}{27}, 1}$.

Problem 3-5

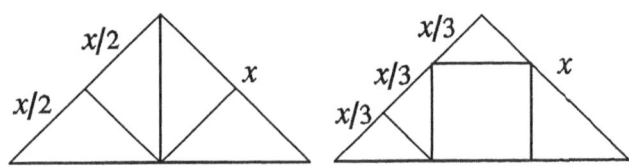

Call the original isosceles right triangle T. In the diagrams, let the length of each leg of T be x. In the left hand diagram, the vertical diagonal of the square partitions T into 4 congruent isosceles right triangles. Since each leg of T is x, the area of the square is $x^2/4$. In the other diagram, draw the segment shown to create 3 congruent isosceles right triangles, each with hypotenuse-length $x\sqrt{2}/3$. That's the length of a side of the square with area $2x^2/9$. Finally, since $x^2/4 = 576$, $x = 48$ and $2x^2/9 = \boxed{512}$ is the area of the smaller square. (The ratio of the squares' areas is 9 to 8.)

A solution without words is shown below.

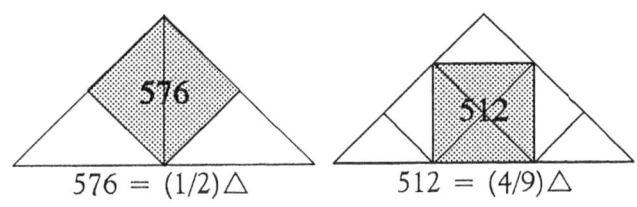

$576 = (1/2)\triangle$ $512 = (4/9)\triangle$

Problem 3-6

The sum of the polynomial's coefficients is $P(1) = 8$, so 8 is an upper bound for the value of any coefficient. In particular, the coefficients are single-digit integers. When written in base 10, $P(10)$ is the sequence of the polynomial's coefficients. For example, If $P(x) = 3x^2 + 4x + 2$, then $P(1) = 9$ and $P(10) = 342$. Here, $P(10) = 2312$, so $P(x) = 2x^3 + 3x^2 + x + 2$. Hence, $P(2) = 2 \times 2^3 + 3 \times 2^2 + 2 + 2 = 16 + 12 + 2 + 2 = \boxed{32}$.

Contests written and compiled by Steven R. Conrad, Daniel Flegler, & Adam Raichel   ©2020 by Mathematics Leagues Inc.

Problem 4-1

We want to add two different primes, but we don't want the sum to be a third prime. Let's add 2 small primes until the sum is composite (but remember that 1 is not a prime): $2+3 = 5$; $2+5 = 7$; $3+5 = \boxed{8}$.

Problem 4-2

The hypotenuse could be either y or $4^2 = 16$. If the hypotenuse is 16 and one leg is $3^2 = 9$, then $y^2 =$ (other leg)$^2 = 16^2 - 9^2 = 256 - 81 = 175$. If 4^2 is a leg, then $y^2 = $ (hypotenuse)$^2 = (3^2)^2 + (4^2)^2 = 81 + 256 = 337$, so y could be $\boxed{\sqrt{175}, \sqrt{337}}$.

Problem 4-3

Perpendiculars to the square from vertices of any side of the octagon not touched by the square form 2 isosceles rt. triangles, each with hypotenuse-length $\sqrt{2}$ (half a side of

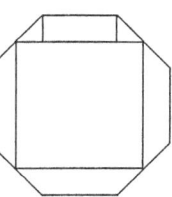

the octagon). The isos. rt. triangles' legs have length 1, so the square's side is $1 + 2\sqrt{2} + 1 = 2 + 2\sqrt{2}$. The perimeter of the square is $\boxed{4(2+2\sqrt{2}) \text{ or } 19.31}$.

Problem 4-4

Since $2^{11} = 2048$ and $2^4 = 16$, $2^{11} - 2^n < 2021$ for $n = 10, 9, 8, 7, 6,$ or 5. That's 6 cases. All the other cases are of the form $2^m - 2^n$ for $m = 10, 9, \ldots, 1$ and $0 \le n \le m-1$: 10 values n for $m = 10$, 9 values n for $m = 9, \ldots$, 1 value of n for $m = 1$. In addition to the first 6 cases noted above, the number of further cases is $10 + 9 + \ldots + 2 + 1 = 55$. The total number of positive integers that can be written as a difference of powers of 2 is $55 + 6 = \boxed{61}$.

[**NOTE:** For integers $a > b \ge 0$ and $m > n \ge 0$, we have $2^m - 2^n = 2^a - 2^b$ if and only if $m = a$ and $n = b$.]

Problem 4-5

Use a regular n–gon as a model, each person at a different vertex. Two people are friends/strangers if they are at vertices connected with a solid/dashed line segment. Three people are friends/strang-

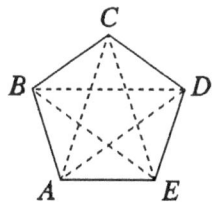

ers if they're at vertices of a triangle with 3 solid/dashed sides. The 5-person pentagon seen at the right shows that, with only 5 people, there need not be 3 friends or 3 strangers since there is no triangle which has 3 solid sides or which has 3 dashed sides each of whose vertices is 1 of the 5 points A, B, C, D, E. The diagram also works for 3 or 4 people. What if there are 6 people? Draw all 15 line segments (some solid, some dashed) that join the 6 points pairwise. Of the 5 segments from any vertex (say, A), at least 3 (say B, C, D) must be the same style, say solid. If any segment connecting $B, C,$ and D is solid, we'd have a solid triangle (with those 2 vertices and A). If not, then $\triangle BCD$ has 3 dashed sides, so the answer is $\boxed{6}$.

[**NOTE:** Ramsey's Theorem generalizes this problem.]

Problem 4-6

For the maximum product, the tens digits must be 9, 8, 7, 6, and 5. For positive numbers with a fixed sum, as their differences decrease, their product increases. In other words, to maximize the product of n numbers, bunch them up closely to their average. Minimize the differences to get $\boxed{90, 81, 72, 63, 54}$.

Here's a formal proof: In any such 2-digit number, the tens digit will be larger than the units digit (otherwise switching them will increase the product). Represent the 2-digit number $90+b$ (with tens digit 9 and ones digit b) as $9b$. We'll show that $b = 0$. Suppose $b \ne 0$. In a maximal product, let the 2-digit number ending in 0 be $a0 = 10a+0$. Since $9b \times a0 = (90+b)(10a+0) = 900a + 10ab < 900a + 90b = (90+0)(10a+b) = 90 \times AB$ (with AB the 2-digit number $10a+b$), we showed that $9b \times a0 < 90 \times AB$. For maximal product, we must have $9b = 90$; so one number is 90. Look at the number with tens digit 8. By an argument like above (but more involved), the ones digit is 1. And so on for the rest. From all this, we get the 5 numbers above.

Contests written and compiled by Steven R. Conrad, Daniel Flegler, & Adam Raichel   ©2021 by Mathematics Leagues Inc.

Problem 5-1

If we add 1 to each of the 2021 positive odd integers, we get the first 2021 positive even integers. Clearly, adding 1 to each increases the average by $\boxed{1}$.

Problem 5-2

$N = (10^{100}-1) - 10^{50} = 999\ldots99 - 10^{50}$. To subtract 10^{50} from a string of 100 9s, we just subtract 1 from one of those 9s. The sum of the resulting 100 digits is $100\times 9 - 1 = \boxed{899}$.

Problem 5-3

The longest *possible* side is 499, the largest integer less than half the perimeter; and, this **IS** the longest side's length since there's a $\triangle$ with sides 2, 499, and $\boxed{499}$.

Problem 5-4

Using the quadratic formula on $x^2+3x+(3-i) = 0$, we get $x = \dfrac{-3 \pm\sqrt{9-4(3-i)}}{2} = \dfrac{-3 \pm\sqrt{-3+4i}}{2}$. To find the square roots of $-3+4i$, we can write $(r+si)^2 = -3+4i$. Equate the real and imaginary parts to get $r^2-s^2 = -3$ and $2rsi = 4i$ from which $(r,s) = (1,2)$ or $(-1,-2)$. Using this in the quadratic formula, we get $x = (-3\pm(1+2i))/2$, so $(a,b) = \boxed{(-2,-1),(-1,1)}$.

Problem 5-5

If the altitude of isosceles $\triangle T$ from its lower left vertex to the opposite leg has length h, then, since 17 is the area of $\triangle$I, $17 = (24/2)h = 12h$. Likewise, the area of $\triangle T$ is $(120/2)h = 60h$. Since $\triangle T$'s area is 5 times that of $\triangle$I, $\triangle T$'s area is $5\times 17 = \boxed{85}$.

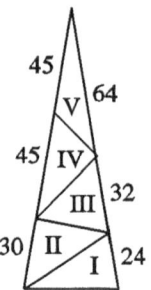

Problem 5-6

Either I can throw more heads than you, or we can throw the same number of heads, or you can throw more heads than I threw. I'll always throw more heads than you or more tails than you. Since I cannot do both, **half the time I get more heads than you.** We'll tie if we both get 0 heads ($P = \frac{1}{4}\times\frac{1}{8}$), 1 head ($P = \frac{2}{4}\times\frac{3}{8}$), or 2 heads ($P = \frac{1}{4}\times\frac{3}{8}$). Thus, the probability of a tie is $\frac{1}{32} + \frac{6}{32} + \frac{3}{32} = \frac{5}{16}$, so the probability p that I'm first to throw more heads is the probability that I win round 1 + the probability we tie round 1 and I win round 2 + the probability we tie rounds 1 and 2, and I win round 3, + the probability we tie rounds 1, 2, and 3, and I win round 4, etc. Therefore,

$$p = \tfrac{1}{2} + \tfrac{5}{16}\times\tfrac{1}{2} + \left(\tfrac{5}{16}\right)^2\times\tfrac{1}{2} + \left(\tfrac{5}{16}\right)^3\times\tfrac{1}{2} + \ldots$$
$$= \tfrac{1}{2} + \tfrac{5}{16}\left(\tfrac{1}{2} + \left(\tfrac{5}{16}\right)\times\tfrac{1}{2} + \left(\tfrac{5}{16}\right)^2\times\tfrac{1}{2} + \ldots\right)$$
$$= \tfrac{1}{2} + \tfrac{5}{16}p, \text{ so } p = \boxed{\tfrac{8}{11}}.$$

Problem 6-1

The fraction will achieve its greatest value when its denominator is minimized. The denominator is minimized when $x = 0$, so the greatest possible value of the fraction is $\frac{4042}{2} = \boxed{2021}$.

Problem 6-2

Extending the common side of the 15-gons creates 2 congruent exterior angles. Since the degree-measure of each exterior angle of a regular 15-gon is $\frac{360}{15} = 24$, the angle x has degree-measure $\boxed{48, \text{ or } 48°}$.

Problem 6-3

Since we're told that $\log_{10}(2000!) = 5735.52\ldots$, it follows that $10^{5735} < 2000! < 10^{5736}$. Integers between 10^n and 10^{n+1} have $n+1$ digits, so the number of digits in the expansion of 2000! is $\boxed{5736}$.

Problem 6-4

If we add 2 to any value of x that satisfies the first equation, we'll have a value of x that satisfies the second equation. Since the solutions of the first equation are 1, 2, 3, the solutions of the second equation are $1+2$, $2+2$, $3+2$, or more simply $\boxed{3, 4, 5}$.

Problem 6-5

Let x represent the number kilometers per liter that the first car gets, and let y represent the capacity of the first car's gas tank, in liters. We are asked to determine the value of xy, in kilometers. We are told that $(x+6)(y-3) = xy$, so $x = 2y-6$. We are also told that $(x-6)(y+6) = xy$, so $x = y+6$. Solving, $y = 12$, $x = 18$, and $xy = \boxed{216}$.

Problem 6-6

Extend the legs of the trapezoid until they meet as shown in the diagram. The 3 triangles in the diagram are all similar, so the ratio of their areas = the square of the ratio of

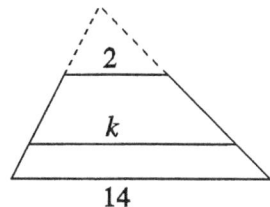

corresponding bases. Therefore, $(2/14)^2 = (1/7)^2 = 1/49 = $ (area of smallest $\triangle$)/(area of biggest $\triangle$). If x is the area of the smallest $\triangle$, then the biggest triangle's area is $49x$. As a consequence, the area of trapezoid T is $49x-x = 48x$ and each of the two smaller trapezoids has area $48x/2 = 24x$. Since the square of the ratio of corresponding side-lengths of similar triangles equals the ratio of the areas of the triangles, it follows that $(2/k)^2 = (x)/(24x+x) = 1/25$, so $k = \boxed{10}$.

Contests written and compiled by Steven R. Conrad, Daniel Flegler, & Adam Raichel   ©2021 by Mathematics Leagues Inc.

Answers & Difficulty Ratings
October, 2016 – March, 2021

Answers

2016-2017		2017-2018		2018-2019	
1-1.	1	1-1.	$\frac{1}{2}$	1-1.	1009
1-2.	4, 7	1-2.	3	1-2.	24
1-3.	81	1-3.	444	1-3.	165°
1-4.	1/24	1-4.	225	1-4.	13
1-5.	100	1-5.	$\frac{2}{3}$	1-5.	9
1-6.	200	1-6.	59	1-6.	$\frac{3}{7}$
2-1.	6	2-1.	-2017	2-1.	4
2-2.	27	2-2.	82	2-2.	0
2-3.	2	2-3.	43	2-3.	8
2-4.	1	2-4.	4, 8	2-4.	16, 80, 400, and 2000
2-5.	$-\frac{1}{2}$	2-5.	33	2-5.	39
2-6.	81π	2-6.	109 336	2-6.	1024
3-1.	4	3-1.	-30	3-1.	1
3-2.	Tuesday	3-2.	$\left(\frac{1}{3},\frac{2}{3}\right)$	3-2.	1
3-3.	3	3-3.	$\frac{1}{9}$	3-3.	2017
3-4.	0	3-4.	504	3-4.	75°
3-5.	$48\sqrt{2}$	3-5.	$\frac{1}{4}$	3-5.	40/9
3-6.	2	3-6.	$13\sqrt{3}/4$	3-6.	41
4-1.	2017	4-1.	8072	4-1.	± 2019
4-2.	4	4-2.	16	4-2.	4
4-3.	**B**	4-3.	4, 5, 10, 20, 40	4-3.	30
4-4.	8π	4-4.	(2,2,5)	4-4.	21
4-5.	–9	4-5.	(1,9), (−1,−14), (23,−1322), (−23,−1323)	4-5.	$\frac{23}{6}$
4-6.	999.5	4-6.	$\frac{1}{3}$	4-6.	87
5-1.	2, 6	5-1.	221	5-1.	891
5-2.	361	5-2.	12108	5-2.	−1
5-3.	$\frac{1}{2017}$	5-3.	$\frac{4}{7}$	5-3.	24
5-4.	$\pm 1, \pm i$	5-4.	$(8-m,n)$	5-4.	71
5-5.	55.44	5-5.	7	5-5.	2
5-6.	8	5-6.	(26,−23)	5-6.	40
6-1.	1	6-1.	2018^2	6-1.	60
6-2.	$\frac{3}{8}$	6-2.	$\frac{11}{6}$	6-2.	252
6-3.	1	6-3.	$\sqrt{2}$	6-3.	10
6-4.	$\frac{1}{3}$	6-4.	(3,2), (6,3)	6-4.	−2
6-5.	$\frac{8}{3}$	6-5.	5	6-5.	$(\pi/6,-\pi/6)$, $(-\pi/6,\pi/6)$
6-6.	102	6-6.	36	6-6.	210

Answers

<table>
<tr><td colspan="2">**2019-2020**</td><td colspan="2">**2020-2021**</td></tr>
<tr><td>1-1.</td><td>2017</td><td>1-1.</td><td>$\frac{1}{2020}$</td></tr>
<tr><td>1-2.</td><td>6</td><td>1-2.</td><td>1</td></tr>
<tr><td>1-3.</td><td>28</td><td>1-3.</td><td>17</td></tr>
<tr><td>1-4.</td><td>660</td><td>1-4.</td><td>1</td></tr>
<tr><td>1-5.</td><td>(31,94), (94,31), (49,86) or (86,49)</td><td>1-5.</td><td>74</td></tr>
<tr><td>1-6.</td><td>45 760</td><td>1-6.</td><td>512</td></tr>
<tr><td>2-1.</td><td>4 076 361</td><td>2-1.</td><td>16</td></tr>
<tr><td>2-2.</td><td>13</td><td>2-2.</td><td>12:30 PM</td></tr>
<tr><td>2-3.</td><td>20</td><td>2-3.</td><td>6062</td></tr>
<tr><td>2-4.</td><td>$121\sqrt{3}$</td><td>2-4.</td><td>8</td></tr>
<tr><td>2-5.</td><td>± 3</td><td>2-5.</td><td>108</td></tr>
<tr><td>2-6.</td><td>$\frac{11+\sqrt{21}}{2}$</td><td>2-6</td><td>11</td></tr>
<tr><td>3-1.</td><td>2019.5</td><td>3-1.</td><td>Dec. 16, 2020</td></tr>
<tr><td>3-2.</td><td>14</td><td>3-2.</td><td>2</td></tr>
<tr><td>3-3.</td><td>5</td><td>3-3.</td><td>10401</td></tr>
<tr><td>3-4.</td><td>1, 5</td><td>3-4.</td><td>$\frac{64}{27}$, 1</td></tr>
<tr><td>3-5.</td><td>$4\sqrt{2}-4$</td><td>3-5.</td><td>512</td></tr>
<tr><td>3-6.</td><td>316 251</td><td>3-6.</td><td>32</td></tr>
<tr><td>4-1.</td><td>1</td><td>4-1.</td><td>8</td></tr>
<tr><td>4-2.</td><td>162</td><td>4-2.</td><td>$\sqrt{175}$, $\sqrt{337}$</td></tr>
<tr><td>4-3.</td><td>21</td><td>4-3.</td><td>$4(2+2\sqrt{2})$ or 19.31</td></tr>
<tr><td>4-4.</td><td>61</td><td>4-4</td><td>61</td></tr>
<tr><td>4-5.</td><td>14</td><td>4-5.</td><td>6</td></tr>
<tr><td>4-6.</td><td>128</td><td>4-6.</td><td>90, 81, 72, 63, 54</td></tr>
<tr><td>5-1.</td><td>$\frac{-11}{6}$</td><td>5-1.</td><td>1</td></tr>
<tr><td>5-2.</td><td>1009</td><td>5-2.</td><td>899</td></tr>
<tr><td>5-3.</td><td>8</td><td>5-3.</td><td>499</td></tr>
<tr><td>5-4.</td><td>13</td><td>5-4.</td><td>$(-2,-1),(-1,1)$</td></tr>
<tr><td>5-5.</td><td>165 & 561</td><td>5-5.</td><td>85</td></tr>
<tr><td>5-6.</td><td>$(-4,-8,-16),(-1,-3,-9),(-1,1,-1,)$</td><td>5-6.</td><td>$\frac{8}{11}$</td></tr>
<tr><td>6-1.</td><td>99990</td><td>6-1.</td><td>2021</td></tr>
<tr><td>6-2.</td><td>84</td><td>6-2.</td><td>48 or 48°</td></tr>
<tr><td>6-3.</td><td>$\frac{1}{64}$</td><td>6-3.</td><td>5736</td></tr>
<tr><td>6-4.</td><td>$2\pi - 4$</td><td>6-4.</td><td>3, 4, 5</td></tr>
<tr><td>6-5.</td><td>$\frac{1}{200}$</td><td>6-5.</td><td>216</td></tr>
<tr><td>6-6.</td><td>$\frac{-1+\sqrt{5}}{2}$</td><td>6-6.</td><td>10</td></tr>
</table>

Difficulty Ratings

(% correct of all reported scores from each participating school)

2016-2017		2017-2018		2018-2019		2019-2020		2020-2021	
1-1.	89%	1-1.	91%	1-1.	96%	1-1.	78%	1-1.	61%
1-2.	84%	1-2.	89%	1-2.	85%	1-2.	82%	1-2.	44%
1-3.	78%	1-3.	57%	1-3.	66%	1-3.	38%	1-3.	56%
1-4.	30%	1-4.	60%	1-4.	51%	1-4.	60%	1-4.	17%
1-5.	31%	1-5.	8%	1-5.	14%	1-5.	51%	1-5.	34%
1-6.	7%	1-6.	7%	1-6.	13%	1-6.	2%	1-6.	6%
2-1.	94%	2-1.	85%	2-1.	59%	2-1.	67%	2-1.	51%
2-2.	35%	2-2.	64%	2-2.	44%	2-2.	58%	2-2.	72%
2-3.	61%	2-3.	70%	2-3.	91%	2-3.	46%	2-3.	66%
2-4.	31%	2-4.	16%	2-4.	23%	2-4.	14%	2-4.	40%
2-5.	23%	2-5.	13%	2-5.	25%	2-5.	12%	2-5.	39%
2-6.	6%	2-6.	5%	2-6.	11%	2-6.	4%	2-6.	24%
3-1.	94%	3-1.	77%	3-1.	55%	3-1.	77%	3-1.	53%
3-2.	77%	3-2.	24%	3-2.	87%	3-2.	49%	3-2.	34%
3-3.	67%	3-3.	8%	3-3.	43%	3-3.	59%	3-3.	65%
3-4.	59%	3-4.	23%	3-4.	34%	3-4.	31%	3-4.	23%
3-5.	28%	3-5.	47%	3-5.	23%	3-5.	20%	3-5.	26%
3-6.	25%	3-6.	4%	3-6.	13%	3-6.	8%	3-6.	22%
4-1.	64%	4-1.	73%	4-1.	85%	4-1.	85%	4-1.	63%
4-2.	70%	4-2.	83%	4-2.	71%	4-2.	57%	4-2.	40%
4-3.	82%	4-3.	63%	4-3.	49%	4-3.	66%	4-3.	40%
4-4.	17%	4-4.	32%	4-4.	31%	4-4.	27%	4-4.	18%
4-5.	23%	4-5.	11%	4-5.	76%	4-5.	22%	4-5.	21%
4-6.	28%	4-6.	14%	4-6.	6%	4-6.	7%	4-6.	51%
5-1.	72%	5-1.	85%	5-1.	91%	5-1.	74%	5-1.	61%
5-2.	87%	5-2.	69%	5-2.	38%	5-2.	47%	5-2.	37%
5-3.	64%	5-3.	59%	5-3.	74%	5-3.	48%	5-3.	50%
5-4.	35%	5-4.	48%	5-4.	74%	5-4.	22%	5-4.	15%
5-5.	41%	5-5.	10%	5-5.	23%	5-5.	68%	5-5.	30%
5-6.	16%	5-6.	5%	5-6.	26%	5-6.	4%	5-6.	9%
6-1.	86%	6-1.	77%	6-1.	88%	6-1.	87%	6-1.	73%
6-2.	61%	6-2.	51%	6-2.	51%	6-2.	85%	6-2.	70%
6-3.	55%	6-3.	35%	6-3.	78%	6-3.	36%	6-3.	48%
6-4.	52%	6-4.	33%	6-4.	50%	6-4.	46%	6-4.	51%
6-5.	35%	6-5.	14%	6-5.	22%	6-5.	36%	6-5.	34%
6-6.	19%	6-6.	13%	6-6.	6%	6-6.	11%	6-6.	33%

Math League Contest Books
4th Grade Through High School Levels
Order books at www.mathleague.com (or use the form below)

Name _____

Address _____

City _____ State _____ Zip _____
                                  (*or Province*) (*or Postal Code*)

Available Titles	# of Copies	Cost
Math Contests—Grades 4, 5, 6	($12.95 per book)	
Volume 1: 1979-80 through 1985-86	_____	_____
Volume 2: 1986-87 through 1990-91	_____	_____
Volume 3: 1991-92 through 1995-96	_____	_____
Volume 4: 1996-97 through 2000-01	_____	_____
Volume 5: 2001-02 through 2005-06	_____	_____
Volume 6: 2006-07 through 2010-11	_____	_____
Volume 7: 2011-12 through 2015-16	_____	_____
Volume 8: 2016-17 through 2020-21	_____	_____
Math Contests—Grades 7 & 8‡	‡(Volumes 3-8 include Algebra 1)	
Volume 1: 1977-78 through 1981-82	_____	_____
Volume 2: 1982-83 through 1990-91	_____	_____
Volume 3: 1991-92 through 1995-96	_____	_____
Volume 4: 1996-97 through 2000-01	_____	_____
Volume 5: 2001-02 through 2005-06	_____	_____
Volume 6: 2006-07 through 2010-11	_____	_____
Volume 7: 2011-12 through 2015-16	_____	_____
Volume 8: 2016-17 through 2020-21	_____	_____
Math Contests—High School		
Volume 1: 1977-78 through 1981-82	_____	_____
Volume 2: 1982-83 through 1990-91	_____	_____
Volume 3: 1991-92 through 1995-96	_____	_____
Volume 4: 1996-97 through 2000-01	_____	_____
Volume 5: 2001-02 through 2005-06	_____	_____
Volume 6: 2006-07 through 2010-11	_____	_____
Volume 7: 2011-12 through 2015-16	_____	_____
Volume 8: 2016-17 through 2020-21	_____	_____
Shipping and Handling	$5 ($10 Canadian)	

Please allow 2-3 weeks for delivery

Total: $_____

☐ Check or Purchase Order Enclosed; **or**

☐ Visa / MasterCard / Discover # _____

☐ Expiration Date _____ Signature _____

Mail your order with payment to:
Math League Press, P.O. Box 17, Tenafly, NJ 07670-0017
or order on the Web at www.mathleague.com
Phone: (201) 568-6328 • Fax: (201) 816-0125